Dark Matter, Dark Energy: The Dark Side of the Universe
Part II

Sean Carroll, Ph.D.

THE TEACHING COMPANY ®

PUBLISHED BY:

THE TEACHING COMPANY
4151 Lafayette Center Drive, Suite 100
Chantilly, Virginia 20151-1232
1-800-TEACH-12
Fax—703-378-3819
www.teach12.com

ISBN 978-1-59803-351-9

Sean Carroll, Ph.D.

Senior Research Associate in Physics, California Institute of Technology

Sean Carroll is a Senior Research Associate in Physics at the California Institute of Technology. He did his undergraduate work at Villanova University and received his Ph.D. in astrophysics from Harvard in 1993. His research involves theoretical physics and astrophysics, with a focus on issues in cosmology, field theory, and gravitation.

Before arriving at Caltech, Dr. Carroll was on the faculty of the Physics Department and the Enrico Fermi Institute of the University of Chicago. Before that, he did postdoctoral research at the Massachusetts Institute of Technology and at the Institute for Theoretical Physics at the University of California, Santa Barbara. Dr. Carroll has made contributions to the theories of topological defects, violations of spacetime symmetries, two-dimensional quantum gravity, anisotropies in the cosmic microwave background, models of brane worlds and extra dimensions, the origin of cosmological magnetic fields, causality violation in general relativity, and quantum supergravity. His current research involves models of dark matter and dark energy, cosmological modifications of Einstein's general relativity, the physics of inflationary cosmology, and the origin of time asymmetry.

While at MIT, Professor Carroll won the Graduate Student Council Teaching Award for his course on general relativity, the lecture notes of which were expanded into the textbook *Spacetime and Geometry: An Introduction to General Relativity*, published in 2003. He has received research grants from NASA, the Department of Energy, and the National Science Foundation, as well as fellowships from the Sloan and Packard foundations. He has been the Malmstrom Lecturer at Hamline University, the Resnick Lecturer at Rensselaer Polytechnic Institute, and a National Science Foundation Distinguished Lecturer. In 2006, he received the Arts and Sciences Alumni Medallion from Villanova University.

Dr. Carroll is active in education and outreach, having taught more than 200 scientific seminars and colloquia and given more than 50 educational and popular talks. He has written for *SEED*, *Sky & Telescope*, *Nature*, *New Scientist*, *The American Scientist*, and

Physics Today. He has also been interviewed on NPR and various local TV and radio programs and has been quoted in *Time*, *U.S. News and World Report*, *Scientific American*, the *New York Times*, *USA Today*, the *Boston Globe*, the *Chicago Tribune*, and numerous other newspapers and magazines.

Table of Contents
Dark Matter, Dark Energy:
The Dark Side of the Universe
Part II

The Dark Side of the Universe:
Dark Matter and Dark Energy

Scope:

For the first time in human history, we know what the universe is made of. A series of remarkable observations at the close of the 20th century enabled cosmologists to discover a complete inventory of the universe: about 5 percent *ordinary matter* (atoms, stars, galaxies), 25 percent *dark matter* (some new particle that hasn't yet been detected here on Earth), and 70 percent *dark energy*. The dark energy is something completely new and unexpected: a kind of energy density that seemingly exists even in completely empty space. Together, the dark matter and dark energy hold the key to the ultimate fate (and, possibly, the origin) of our universe.

This course will examine why we think that dark matter and dark energy must exist and what they might be telling us about the mysteries of fundamental physics. We will start with the picture of our universe that cosmologists constructed over the course of the 20th century. Our universe is big: It consists of billions of galaxies, each containing billions of stars, scattered nearly uniformly throughout space. And it's getting bigger: Galaxies are moving apart from each other as the space between them expands. The entire picture is accurately described by general relativity, Einstein's dramatic theory of spacetime and gravitation.

Gravity rules the cosmos on very large scales, and we can use it to probe the ingredients that compose the universe. Even if we don't see something directly, it will still give rise to a gravitational field. Following this logic, the motions of galaxies imply the need for more matter than we see directly: We call the extra stuff *dark matter*. We might suspect that the dark matter is just ordinary matter that isn't emitting or absorbing light, but that possibility can be ruled out. Particle physics as we know it is summarized in the Standard Model, a triumphant theory that fits just about every experiment performed here on Earth. But there are no particles in the Standard Model that could describe the dark matter—it must be something new and undiscovered.

The biggest reason why dark matter can't be any particle we know comes from our picture of the very early universe. Tracing backwards in time toward the Big Bang, we infer that the universe went through a series of transitions during which particles came together or were torn apart, such as in the first few minutes of its history, when free protons and neutrons combined to form helium and other elements. We see the remnants of these transitions today, and they imply strict bounds on the amounts of ordinary matter present in the universe. We also see the cosmic microwave background, relic radiation from when the universe first became transparent. The tiny fluctuations in the temperature of this radiation are tied to the composition of the universe and provide additional evidence that dark matter is something new.

Even newer is dark energy—something that isn't made of particles at all but is smoothly distributed through space and slowly evolving (if at all) through time. The dark energy has dramatic effects on the dynamics of the universe: It affects the overall curvature of space and causes the expansion rate of the universe to accelerate. These effects have been observed, leaving little doubt that dark energy does exist. But what is it?

Fundamental physics offers a few possible candidates for the origin of dark energy, but none makes perfect sense to us. The simplest is vacuum energy, or the cosmological constant—an absolutely constant energy density inherent in spacetime itself. The idea of vacuum energy is quite natural, but its magnitude is not; if vacuum energy exists at all, it should be much greater than what we observe. Alternatively, cosmologists have imagined that dark energy arises from some new dynamical field dubbed *quintessence.* This possibility has the attractive feature of being experimentally testable; if dark energy is varying with time, future experiments can reveal it.

Despite multiple lines of supporting evidence, there is the more dramatic possibility that dark matter and dark energy actually do not exist, but that instead, we are being tricked by the behavior of gravity. Einstein's theory of general relativity has been very successful, but there remains the possibility that it is incomplete on the very largest distance scales. Physicists have been busily inventing alternative theories, but they are tightly constrained by experimental tests. They have also been trying to reconcile general relativity with the formalism of quantum mechanics, efforts that have given rise to the ideas of string theory. Although experiments to verify string theory haven't yet been devised, string theory holds out the intriguing possibility that the part of the universe we see is highly atypical—that we live in a tiny corner of a multiverse in which the local laws of physics are determined by their hospitability toward life as much as by deep, underlying principles.

Clearly, the challenges raised by dark matter and dark energy are pointing physicists and astronomers far beyond their current theories. Only an ambitious program of future observations and experiments will give us the necessary clues to build a comprehensive model that explains all of the dark side of the universe.

Lecture Thirteen
WIMPs and Supersymmetry

Scope:

If the dark matter isn't in the Standard Model, it's a new particle. That particle may be heavy and interact through the weak nuclear force—thus, *weakly interacting massive particles*, or WIMPs. It's up to particle physics to provide candidates for what exactly that might be. In this lecture, we consider specific models for the dark matter. There are many such models, but one stands out as especially promising: supersymmetric particles. Supersymmetry is the bold idea of a hidden symmetry between bosons and fermions. It requires that there are twice as many kinds of fundamental particles than we have heretofore observed—a boson for every known fermion and vice versa. The lightest new supersymmetric particle can be stable and is an excellent candidate for a WIMP. Other candidates include axions, sterile neutrinos, and Kaluza-Klein particles.

Outline

I. Weakly Interacting Massive Particles are an ideal dark matter candidate.

    A. Measurements of the baryon density from nucleosynthesis and the microwave background drive us to the idea that the dark matter is a kind of particle.

        1. The only possibility in the Standard Model are the neutrinos, which are both dark (they don't interact with photons) and have a small mass. However, the details of this idea don't quite work.

        2. Although we now believe that neutrinos have mass, those masses should be very small.

        3. When neutrinos were produced in the early universe, they were *hot dark matter*—moving close to the speed of light. That's bad news for the formation of structure, because the hot dark matter tends to smooth out primordial density perturbations before they can grow into galaxies.

4. We believe that *cold dark matter* must be dominant in the universe, which speaks against the idea that neutrinos are the culprit.

B. There are, thus, no good candidates for dark matter in the Standard Model.

1. We have to invent a new kind of particle—one that is massive (cold), neutral (dark), and stable enough to last for the lifetime of the universe.

2. One good idea is that the relevant particle interacts only through the weak nuclear force; such particles are called *weakly interacting massive particles*, or WIMPs for short.

3. Physicists are hot on the trail of WIMPs, both experimentally and theoretically.

II. The hypothetical concept of supersymmetry suggests an origin for WIMPs.

A. *Supersymmetry* is a type of symmetry that relates bosons to fermions.

1. For every kind of boson, there will be a kind of fermion with identical properties and vice versa.

2. As yet, this idea is completely conjectural—no evidence for supersymmetry has been found in nature. But it's a compelling idea, both theoretically and experimentally.

B. Naively, supersymmetry predicts that the electron, which is a fermion with charge −1, should have a bosonic partner with the same charge and the same mass.

1. Physicists have a cute naming convention for such new particles, collectively known as *superpartners*: The superpartners of ordinary fermions get an *s* tacked on to the front, while the superpartners of ordinary bosons get *-ino* tacked onto the end.

2. Thus, the superpartners of electrons are *selectrons*, and the partners of quarks are *squarks*; the partners of photons are *photinos* and so on.

C. It should be very easy to detect selectrons, but we haven't done so. Thus, if supersymmetry is real, it must be somehow hidden.

1. That's actually not so hard to do; *symmetry breaking* is a common phenomenon in particle physics and would work to give all the new supersymmetric particles much larger masses than their ordinary partners.

2. That means that there might be a large number of new particles to be discovered, perhaps at the next generation of particle accelerators.

D. In order to make supersymmetry work, individual superpartners can decay into other superpartners but not exclusively into ordinary particles. This means that the lightest supersymmetric particle (LSP) is both heavy and stable—an ideal dark matter candidate!

1. Such particles would most likely feel the weak nuclear force, which means that they are WIMPs.

2. Indeed, supersymmetric WIMPs are the single leading candidate for the dark matter particles.

III. It's not enough to have these good ideas; we need to test them. A variety of experiments are currently running that would allow us to detect dark matter particles directly.

A. *Direct detection* is the attempt to build an experiment here on Earth that is sensitive to the presence of ambient dark matter particles.

1. In most models, dark matter particles pass through your body all the time; you just don't notice because they don't interact very strongly.

2. But if you collect enough material and your detectors are sufficiently sensitive, you might be able to notice a dark matter particle scattering in your experiment.

B. *Indirect detection* is the idea of looking for photons created when two dark matter particles collide.

1. If the dark matter is heavy, these photons could be very high energy, and new gamma-ray telescopes are being constructed to look for them.

2. These telescopes will point at the centers of galaxies, where the concentration of dark matter should be especially high.

C. Finally, we can simply make the dark matter particles ourselves.

    1. Supersymmetry, in particular, is a major target of experiments at particle accelerators; other candidates, such as axions, require specialized detectors.

    2. It's hard to predict when we might expect definitive results from such experiments, but the next 5 to 10 years could teach us a great deal about the nature of dark matter.

IV. Supersymmetry is the leading idea for dark matter, but it's not the only one. We could have WIMPs that arise from some other mechanism.

    A. For example, Kaluza-Klein theory imagines that we have extra curled-up dimensions of space (as in string theory, which we'll discuss later). If the size of those dimensions is just right, particles with motion in the extra dimensions would look to us as if they have extra mass, and their conserved momentum would render them stable.

    B. On the other hand, we could have cold dark matter particles that aren't WIMPs at all.

        1. For example, we could have new *sterile* neutrinos, that don't feel the weak interactions.

        2. Or we could have *axions*, hypothetical particles that are very light and created via a special mechanism.

            a. According to this hypothesis, axions arise when a certain kind of field is "frozen" in the early universe and "melts" when the expansion rate dips below a certain value. Even though they are very light, they are not moving rapidly when they are created; thus, they are cold dark matter.

            b. Axions require their own special experimental searches, which are actively underway.

Recommended Reading:

Guth, *The Inflationary Universe*, chapter 14.

Nicolson, *The Dark Side of the Universe*, chapter 7.

Randall, *Warped Passages*, chapter 13.

Weinberg, *Dreams of a Final Theory*, chapter 5.

Questions to Consider:

1. Explain why a good dark matter candidate should or should not have the following properties and which properties don't matter.
   a. Electrically charged or neutral.
   b. Strongly interacting or not.
   c. Stable or unstable.
   d. Massive or massless.
   e. Bosonic or fermionic.

2. Imagine that some heavy particle was produced in the early universe but decayed into some lighter (but still massive) particles after 1 million years, so that the dark matter changed its identity. Is there any way we could test this idea?

3. Is it possible that the dark matter is a complicated system of particles that interact with each other and form dark stars and planets and people?

Lecture Thirteen—Transcript
WIMPs and Supersymmetry

By now, we have absolutely squelched any remaining suspicion that you might have in your mind that the dark matter, evidence for which we found in the dynamics, the motions of galaxies and clusters, could somehow possibly be ordinary matter that was somehow hidden. It can't be gas or dust because gas and dust would fall into clusters of galaxies, heat up, and we would be able to see it in X-rays. It also very probably cannot be ordinary stars, brown dwarfs, white dwarfs, neutron stars, or black holes because those lead to microlensing events, which we've searched for and can't find. Finally, the real reason we know that ordinary matter in some hidden form can't be the dark matter that we have evidence for is both Big Bang nucleosynthesis and the cosmic microwave background put a very tight constraint on how much ordinary matter there is in the universe. It's only 5 percent of the critical density, but the dark matter is something more like 30 percent. We have to, in other words, turn to some unordinary kind of matter, some new kind of particle.

In this lecture, we're going to get serious and start saying, what kind of particle could be the dark matter? We have to go beyond the particles that we have already established exist in the Standard Model of Particle Physics. What are the requirements for a new dark matter particle? Most obviously, it must be dark. The other important requirement is that it must be cold. So, dark means not only is it hard to see, but it doesn't interact very much.

When you look at a galaxy, the shiny part of a galaxy, the part that is easily visible from stars comes from the ordinary matter, of course. The reason why the galaxy is able to contract is because ordinary matter interacts. Ordinary matter, in the process of collapsing under its own gravitational field, will get stuck. When one little bit of ordinary matter comes in contact with another one, it can cool off and settle down into the bottom.

Dark matter seems to be distributed in a big, puffy halo around the galaxy. It's not condensed right in the middle like the visible matter is. The explanation for this is very easy to come up with. It's that the dark matter particles just pass through each other and go right out.

So, dark matter needs not only to be dark; it needs to be not interacting with itself in any obvious way.

The other requirement for the dark matter particle is it should be cold. In other words, even if the dark matter existed and had a small mass, but it was moving at a very fast velocity, if it was moving close to the speed of light in the early universe, then when it tried to collapse, it would go right outside and keep going. It would not oscillate back and forth like we want a real good dark matter particle to do.

We're trying to invent a new kind of particle that is dark and is cold. Let's just check that there is no such particle lying around in the Standard Model. But, which particles in the Standard Model are dark? Of course, we don't want the ones that decay. Neutrons would not be good examples of dark matter particles because they decay away. You need something that will stick around. The only neutral particle in the Standard Model that is stable is the neutrino. Neutrinos are a very obvious place to look for a dark matter candidate. However, as it turns out, when you go through the details, because neutrinos are so light, they are not cold. When neutrinos decouple, when neutrinos freeze out from the primordial plasma at very early times, they are moving very, very fast, so they qualify as *hot dark matter*.

If the dark matter we observe is hot, then structure on small scales in the universe doesn't form. You don't make galaxies in a universe that is mostly hot dark matter because, whereas the stuff would want to collapse, the neutrinos are moving out very, very quickly and tend to smooth everything out. Hot dark matter is actually quite strongly ruled out as a possible way for the dark matter to behave, so much so that these days we turn it around. We use cosmology to place a constraint on the mass of the neutrino. If the neutrino mass is too big, it would be the dark matter. But, we know it can't be the dark matter because the dark matter is not hot. That's a good way to get an upper limit on how big the neutrino mass could be.

That's the only possibility in the Standard Model of Particle Physics. We need to turn to particles beyond the Standard Model. Let's think of what the requirements are on those particles. Of course, they must be dark and must be cold. But, in Lecture 9, we gave an explanation for how to calculate the relic abundances of particles that are left over from the early universe. If a particle interacts strongly in the early universe, then the particle and its anti-particle will all annihilate away. By the end of the universe's history, by today, we won't have anything left over. What you need for a particle to be left over in sufficient amounts to be the dark matter is a particle that does not annihilate very strongly, a particle that is weakly interacting.

When we say weakly interacting, you might think—and it might be perfectly sensible—not interacting very much. It doesn't interact very often when two particles come together and that's true. But, now, let's actually plug in the numbers. Let's do our calculation of what kinds of particles give rise to the right abundance of a particle today to be the dark matter and ask how often should particles annihilate in order for them to be good dark matter candidates. The answer is that the right rate of interaction for a new particle to be the dark matter candidate is exactly what it would have if it interacted through the weak nuclear force. When we say weakly interacting particles are good dark matter candidates, we don't simply mean particles that don't interact very much. We mean particles that interact via W and Z bosons, particles that interact via the weak interactions of the Standard Model. It's not the only possible way to get dark matter, but it's suggestive. It's telling us that if we invent a new particle that is stable, that is not interacting through electromagnetism or the strong interactions, but does interact with the weak interactions, it is a natural dark matter candidate. The name attached to such a candidate is a weakly interacting massive particle, or a WIMP, as opposed to the MACHOs, which are the compact collections of ordinary matter.

We want to make WIMP candidates. We want to make new examples of particle physics that give us weakly interacting massive particles that are stable and could be the dark matter. For most of this lecture, what I'm going to do is talk about one specific example of a particle physics model that naturally leads to a candidate WIMP dark matter particle. This specific example is called *supersymmetry*. Really, what you should be getting out of this lecture is not the details of supersymmetry as a candidate way to get dark matter; it's just an example of the kinds of thought processes that physicists go through when they are trying to invent new particles. The point is that you don't want to just say, "Well, yes, there must be some particle and it's the dark matter."

Particles have interactions. Particles come with different properties. We know a lot already about the particle physics of the Standard Model, so when you're inventing a new particle, it has to fit in somehow. The best dark matter candidates will be those that have some natural interpretation in terms of particle physics all by themselves, even if we didn't know that there was such a thing as dark matter. Supersymmetry is an excellent example of that, and that's why it's worth going into it in some detail.

Supersymmetry is a hypothetical idea, which invents a new symmetry onto the symmetries we already know about in the Standard Model. We already mentioned that the Standard Model of Particle Physics is characterized by a great amount of symmetry. For example, the most obvious thing that you see in the Standard Model is that when you look at the fermions, they come in three generations. You have the up and down quark in a doublet, the electron and its neutrino in a doublet, and all by themselves, those four particles form a self-contained set. Then, you have another four particles, which repeat that pattern—two quarks and two leptons, the charmed and strange quark, and the muon and the muon neutrino. Then, it happens yet again with the bottom quark and the top quark, the tau and the tau neutrino.

The fact that you have a similar structure repeating itself over again is an example of symmetry. At a deeper level, symmetries are responsible for the forces between particles, the strong nuclear force, the weak nuclear force, and the electromagnetic force. Supersymmetry is a very specific kind of symmetry, a different kind of symmetry than we have ever seen in the Standard Model. It's a symmetry that relates bosons to fermions. Bosons are the force particles, the particles that can pile on top of each other to give rise to electromagnetic fields, gravitational fields, and so forth. Fermions are the matter particles. They are the particles that take up space. They seem very different from each other. Remember, there is a thing called spin, and bosons always have an amount of spin that is zero, one, two, etc. Fermions always have an amount of spin that is one-half, three-halves, five-halves, etc. So, supersymmetry is a speculative idea. It's saying that somehow there could be a symmetry relating particles of different amounts of spin. If that is true, the nice thing is that you can have a light supersymmetric particle, which is a perfect dark matter candidate. Such a particle is sometimes called the *neutralino*, and that's what we want to explore here.

If supersymmetry existed, for every kind of fermion particle, there would be a boson particle with the same kinds of charges and the same mass, and vice versa. For every boson, there would be a corresponding fermion. For example, you have a bosonic particle that has a certain mass, a certain *electric charge*, and a certain interaction under the weak nuclear force and the strong nuclear force. If supersymmetry existed, there would be a fermion with the same mass, the same electric charge, and the same weak and strong nuclear force interactions, but a different spin. Now you can look in the Standard Model. There are certainly particles of both kinds of spins in the Standard Model. There are bosons and fermions, but they certainly don't match up. You can't take the bosons of the Standard Model and say, well, for this boson, this fermion is obviously its super partner. It doesn't quite go like that.

You could also imagine that there are new particles; that there are particles that we don't see in the Standard Model yet. But, those could be the *superpartners* of the particles that we know and love.

For example, the electron would have a partner that was a boson and it would be called the *selectron*. It would have a charge of minus one, just like the electron does. It would have a lepton number of plus one, and it would have a quark number of zero. It would not interact with the strong interactions, but it would interact with the weak interactions. But, if supersymmetry were exactly right, that selectron would have exactly the same mass as the electron and that's clearly not true. If there were a bosonic particle with the electric charge of the electron and the same mass of the electron, we would have noticed it long ago.

So, somehow, we have to invent a whole new bunch of particles, a bunch of particles we have never yet seen. What we have done is we've given them names. That's a good first start. If we haven't actually found them yet, at least we can come up with their names. The new particles that we need to imagine, if we're going to believe that supersymmetry is right, are the fermionic partners of the existing bosons and the bosonic partners of the existing fermions. The bosons that are the partners of the existing fermions are given names that are derived by tacking an "s" onto the beginning of the name of the fermion. So, we have different kinds of quarks, for example. Their bosonic superpartners are called *squarks*. We have different kinds of leptons. Their bosonic superpartners are called *sleptons*. You have the electron and its partner the selectron, the neutrino and its partner the *sneutrino*, etc. You can have great fun with this if you go very far.

For the existing bosons, they have fermionic partners that are given names that you get by tacking the suffix "ino" at the end of the particle name. You have, for example, the photon, which is a boson. It has a fermionic super partner called the *photino*. The graviton is a boson; there is the *gravatino* fermion. The Higgs boson we think exists has a partner called the *Higgsino*, and so forth.

You get all sorts of particles. If supersymmetry is right, given the fact that there is no way to match up the particles that we already observe, supersymmetry is hypothesizing that we have double the number of particles in the real world than we have actually observed in the Standard Model of Particle Physics. It is a very economical idea, in the sense that it's just one little idea. Just say there is a symmetry between bosons and fermions; it's easy to say.

But, it becomes very prolific in terms of what kinds of particles it predicts. It's not just one more particle tacked onto the Standard Model; it's doubling the number. Now, you might ask, is it worth doing that? Why are we contemplating doubling the number of particles that we have in nature? Dark matter is a nice benefit from inventing supersymmetry, but it's not the primary motivation. In fact, the very first reason why supersymmetry was invented is because it's predicted by string theory. String theory, which is something we'll discuss later in the course, is a hypothetical way of quantizing gravity, and it turns out that string theory only works if you have supersymmetry. That's why supersymmetry was invented. But, then, it was recognized that even if string theory isn't right, supersymmetry by itself is not only aesthetically pleasing, it's not only a very nice and elegant symmetry of nature, it also solves some naturalness problems that exist in the Standard Model.

The foremost one is called the hierarchy problem. We're not going to go into great detail about the hierarchy problem, but it's basically the fact that the different mass scales of particle physics are very, very different from each other. That is the kind of thing that particle physicists don't like. They don't like things to be very, very different from each other unless there is some good reason. So, the mass of the Higgs boson in the Standard Model of Particle Physics, which is the thing that sets the mass scale for all the massive particles in the Standard Model, is somehow very, very different than the high energy mass scales we're familiar with from gravity, from the *Planck scale*, or from grand unification theories.

Why is the Higgs boson so much lighter than the particles that we would expect to exist at very high energies? That's called the hierarchy problem. It turns out that, in supersymmetry, there is a natural explanation for the hierarchy problem. That's why most particle physicists like supersymmetry as a candidate for physics beyond the Standard Model. The problem, of course, is that we don't see selectrons, we don't see squarks, etc. Somehow, they must be hidden from us. In particular, they must weigh a lot more. They must have a much bigger mass than we would naturally expect. How do you do that? The answer is that this symmetry you're inventing, supersymmetry, must somehow be hidden from our immediate view.

The idea that symmetries are hidden is a very familiar one in particle physics. In fact, it's the correct way to think about the weak interactions. We mentioned that the W and Z bosons that carry the weak interactions are very heavy particles, unlike photons, gluons, and the graviton, which are very light particles. Why is that? In their natural state, the W and Z bosons that carry the weak interactions would be massless, but it turns out that the symmetry associated with those particles is broken by empty space—in fact, by the Higgs field, that's why the Higgs field must exist. The role of the Higgs field is to break the symmetry of the weak interactions and give mass to the W particles and the Z particles.

This sounds like maybe cheating. You're trying to sort of hide something in a broken symmetry to explain things you don't otherwise understand. But, the idea that properties of particles change depending on where they are or the medium through which they move is very natural. For example, light does not always travel at the speed of light. The speed of actual light rays is different in air, or in water, or a glass, than it would be in empty space. That's because the medium through which the light is traveling has properties all by itself. So, when we speak of the speed of light, we really mean the speed of light in a vacuum. But, in stuff, the speed of light can be very different.

Similarly, the idea of a broken symmetry is that some field pervades empty space. Modern particle physics says that even in empty space, there is a Higgs field, something that has a non-zero value. The reason why W and Z bosons have mass is that they are traveling through this Higgs field. That's a very successful idea. The particles that were predicted by this idea have largely been discovered and Nobel Prizes have been given out. We just want to do the same kind of thing, but now we want to do it with supersymmetry. We want to break supersymmetry. There must be some mechanism that breaks that symmetry at a deep level.

If that happens, then all of the superpartners of the particles in the Standard Model become heavy. Basically, you take a whole bunch of particles that had the same masses as the particles we observe, but you lift their masses by a large amount. You raise them by breaking the symmetry and you end up with a whole bunch of particles, all of which are very, very heavy, at least a thousand times the mass of the proton. The reason why we haven't discovered supersymmetry yet in this scenario is that it's just too hard to get there. It's just out of our reach, although we're trying to do it right now.

An interesting wrinkle of this possibility is that there can be a new conserved quantity associated with supersymmetry. Remember that there are things called quark number, lepton number, and electric charge, which have a quantity that can neither be created nor destroyed. An electrically charged particle cannot decay into neutral particles. What if superness is also a conserved quantity? What if whether or not you are in the Standard Model or a super partner of the Standard Model is a conserved quantity? Then, the superpartners of the Standard Model would not be able to decay into the partners we see, into the actual particles that we know exist in the Standard Model; therefore, the lightest supersymmetric particle would have to be stable. There would be nothing for it to decay into.

What we're saying is that there would be a kind of particle that would be stable because it carries some conserved quantity. It is heavy. It's a thousand times greater than the proton, so we haven't seen it yet, and it can very plausibly be weakly interacting. These are particles that are part of the supersymmetric Standard Model. They are not completely separate. So, you could have particles, like the partner of the Higgs boson would be Higgsino. That would be an electrically neutral particle, but it would feel the weak interactions.

Likewise, the *zino*, the partner of the Z boson, or even the photino, the partner of the photon, these would all be massive, possibly stable particles, any one of which is a candidate for the lightest supersymmetric particle. Any one of them would make an excellent WIMP. They are all weakly interacting massive particles. We don't know which one is the right one, because we don't know which one is lightest. Under different scenarios of supersymmetry, different particles are going to get different masses, so that's one of the things we need to figure out by taking data. We're not going to know, until we do experiments and find the superpartners, which one of them is actually the lightest—which one is the LSP, the lightest supersymmetry particle, or the neutralino.

The reason why we like this theory is because supersymmetry was not invented to give us a dark matter candidate. It was invented for other reasons, but lo and behold, a perfect dark matter candidate pops out. It's easy in supersymmetry to get particles that are stable and weakly interacting and massive. What you want to go do is test this idea. You want to go look for these particles. There are various ways to go look for them. The one that is sort of the most promising over the next few years is called *direct detection*, which is to build an experiment that will actually find a dark matter particle directly. The problem is dark matter particles, by construction, are weakly interacting. They don't interact very much. What you have to do is very similar to what has already successfully been done with neutrinos. You need to build a detector that is deep underground, shield it from the noise that we're subjected to here on the surface of the earth, and is very, very sensitive to particles coming in and lightly glancing off of an atomic nucleus in the detector.

We've already done this for neutrinos, but neutrinos are, number one, at a very different energy scale than the dark matter particles are. Number two, there is a shining beacon in the sky that emits neutrinos—namely, the sun. For dark matter particles, what we're looking for is a background of dark matter particles. We don't have a shining source to look at, so it's harder to do. But, there are a number of experiments going on right now that are actively trying to do exactly this. It's very plausible that over the next couple of years, there will be a headline in the newspapers in the morning saying, "Scientists have directly detected the dark matter of the universe." But, maybe they won't. You don't know. That is a hard thing to do. In different models, it becomes very easy or very difficult. So, we're trying other ways.

One other way is a very clever idea, taking advantage of the fact that not only do you have dark matter particles; you have dark matter anti-particles. The reason why the dark matter particles have a certain density is because they have stopped annihilating because the universe has expanded. They have both particles and their anti-particles in the dark matter. They just don't annihilate because they don't interact with each other. There aren't that many of them around. There aren't that many dark matter particles per cubic centimeter. Much like neutrinos, they can pass right through each other very easily. You need a very high density of dark matter particles before you'd begin to see them annihilate.

But, there are places in the universe where the density of dark matter particles might be very high. At the center of a galaxy, or the center of clusters of galaxies, the dark matter particles that were very spread out will collapse. They will contract and they will gather in the same place, and there you will begin to see dark matter particles annihilating with dark matter anti-particles. When that happens, they are going to give off radiation. They are going to give off high energy photons and, in most models, what they will give off are *gamma rays*. Gamma rays are hard for us to observe here on earth because if a gamma ray from the sky comes to us here, it will be absorbed by the atmosphere.

So, NASA and the Department of Energy are building a new satellite called GLAST that will be launched in 2007. GLAST is going to be looking for gamma rays at the centers of galaxies and the centers of clusters of galaxies. What that means is we'll be indirectly detecting dark matter. If we see a signal of gamma rays from a very specific source at a very specific energy, that is exactly what you'd expect if a particle and its anti-particle were annihilating. Again, this may or may not happen in different models. We have to go out there and do the experiment to see if nature is being nice to us in this way.

Finally, we have perhaps the most direct method of all, which is that, forgetting about the dark matter that is surrounding us, let's go and make our own dark matter. This is what particle physicists are paid to do, to collide energetic particles together and to make new ones. This is what we're trying to do right at this moment, and we're building better particle accelerators to do even better. The reason why it's not a surprise that we haven't yet made dark matter particles is because it's hard to notice even if we do. Dark matter particles are weakly interacting, so they are very hard to make. They are neutral, so they are very hard to detect once you've made them. In other words, we might be making dark matter particles all the time in current particles accelerators, but we just don't have enough data to be sure that that is what is going on.

As of 2007, the most high-energy collisions that we can produce here on earth are being produced at the Fermi National Accelerator Laboratory, or Fermilab, outside Chicago. They are making high-energy collisions at about a thousand times the mass of a proton at about the place you would just expect to see superpartners beginning to be created. The problem is, since it's just at the edge, you might make one or two and you would never know. So, we're crossing our fingers. It's certainly possible that the Tevatron, the accelerator of Fermilab, could make supersymmetric particles, but we can't guarantee it. Therefore, we're trying to build even better accelerators. The Large Hadron Collider is being built right now at CERN, a European particle accelerator outside Geneva. It is scheduled to turn on in late 2007, but it will take at least a year for the energies to ramp up. Once they do ramp up, the energies at the Large Hadron Collider, the LHC, will be ten times more than the energies at Fermilab. We'll be able to create vastly new numbers of particles at the LHC that we could only barely hint at, at the Tevatron.

It's, again, very plausible that just a couple of years after the accelerator turns on, it will be awash in supersymmetric particles. The problem there will be there will be too many of them. We'll have to figure out what is going on. This is not going to be an overnight project, but there is going to be a lot of excitement involved when we discover new particles at the Large Hadron Collider, in trying to make sense of them, trying to figure out how these new particles fit into particle physics, whether or not one of them could be the dark matter.

It's also possible, of course, that there is dark matter, that the dark matter is some neutral particle that does not interact very strongly, but that does not come from supersymmetry. So, there are candidate particles for dark matter, both that qualify as WIMPs, weakly interacting massive particles, and other kinds of dark matter particles.

I'll give you one example of a particle that is a different way to get a WIMP, a particle that is neutral, stable, and feels the weak interactions. That's something called the lightest Kaluza-Klein particle. This is an idea that has nothing to do with supersymmetry. It says that there are extra dimensions of space. There are tiny directions that you can go in space that are curled up into little balls, so that you can't see them. This is something that we'll talk about in some detail in the last few lectures of this course. The point is, if you have tiny curled up dimensions, then every particle that we already know and love—electrons, photons, and what have you—have an infinite number of partner particles that correspond to particles that are moving in the extra dimensions with different amounts of momentum.

Because of quantum mechanics, these different amounts of momentum are not arbitrary; they are quantized. So, there is a minimum amount of extra energy that a particle can have from spinning around in the extra dimensions that would show up to us as something called the lightest Kaluza-Klein particle. It could very easily be weakly interacting and massive. That's another very promising candidate for a WIMP and, therefore, for the dark matter.

Then, there are also particles that are not weakly interacting at all. In other words, they are the dark matter, but they don't feel the W and Z bosons. One example is neutrinos, but *sterile* neutrinos. A sterile neutrino is exactly a neutrino that doesn't feel the weak interactions. That's what the word sterile means, in this particular context. So, you can invent new kinds of neutrinos. Neutrinos already don't feel the electromagnetic force or the strong nuclear force. These are the kinds of neutrinos that don't feel *electromagnetism*, the strong force, or the weak force. All they feel is gravity and they can occasionally interact with other kinds of neutrinos. People are making models of massive sterile neutrinos, calculating how many you can make in the early universe, and it's easy to get the right abundance to be the dark matter.

The only reason why this kind of model is not as popular as supersymmetry is that you don't get a lot extra out of it. The bonuses that you get from supersymmetry are quite considerable just from the particle physics perspective. Sterile neutrinos help you a little bit, but we don't know whether they are part of some bigger picture yet.

Finally, let me mention *axions*. Axions are perhaps the second leading candidate for dark matter particles after supersymmetric particles. But, axions are really completely different in conception than the supersymmetric, the LSP, or the neutralino would be. Axions are bosons, whereas the supersymmetric particles that would be the dark matter are fermions. The supersymmetric particles are very heavy, a thousand times the mass of the proton. Axions are very light. Axions have the same kind of mass that a neutrino has. They are very, very low mass particles, which ordinarily you would expect would be fast moving. Neutrinos can't be the dark matter because they are so light that they are moving very fast, and they do not make good dark matter candidates. Why is it that axions, which are very, very light, can nevertheless be *cold dark matter*?

The answer is the axion is created by a very different mechanism than neutrinos are or WIMPs are. The axions, in the models that people write down, were never interacting with the rest of the particles in the plasma of the very early universe. They were never heated up by interacting with the rest of the stuff in the primordial soup. Instead, there was a kind of field, an axion field that didn't change. It was just stuck there and it contained energy. This energy was just constant. It wasn't going away until a phase transition happened and this field melted. When this field melted, it went from being a constant amount of energy per cubic centimeter to a bunch of axions. The field melts into a bunch of axions with zero velocity, a completely different mechanism than you get from making weakly interacting massive particles or neutrinos. It turns out that there are enough free parameters in the model to make these kinds of axions from a melting field with exactly the right kinds of density to be dark matter.

This is good news and bad news. It's good news because it's a completely different way to get dark matter particles. The bad news is that, therefore, the ways to go and look for axions are completely different also. The kinds of experiments we're doing to try to find WIMPs in underground detectors, in the sky, and in the laboratory, have a set of corresponding experiments we would like to do for axions, but they are different experiments. People are still doing those experiments. We're very hopeful that we'll find either WIMPs or axions. We might even get especially lucky. The best universe, if you're a theoretical physicist, would be one in which half of the dark matter is supersymmetric particles and half of it is axions. We will actually have to do the experiments to see whether nature is so kind to us as that.

Lecture Fourteen
The Accelerating Universe

Scope:

How do we know that we've accounted for all the matter in the universe? One way is to test how the expansion rate changes with time, which is related to the galaxies, clusters, and other "stuff" in the universe by general relativity. In the 1990s, two groups undertook to measure the evolution of the expansion rate, using supernovae as standard candles. To everyone's surprise, they found that the expansion was speeding up rather than slowing down—an accelerating universe. Such behavior can't be explained by any kind of matter, ordinary or dark, and thus, suggested the existence of an entirely new component: dark energy. A new *concordance model* of cosmology was constructed, in which about 70 percent of the energy in the universe is dark energy, 25 percent is in dark matter, and only 5 percent is in ordinary matter.

Outline

I. We can "weigh" the entire universe.

    A. By observing the dynamics of galaxies and clusters, we have established the need for dark matter. But how can we be sure that there isn't still more "stuff" out there in the universe, perhaps in between the galaxies and clusters?

        1. In particular, the observed matter density is only about 30 percent of the *critical density*—the density needed, according to the Friedmann equation, to accommodate a spatially flat universe.

        2. The observed value is close enough to the critical density—without equaling the critical density—that it suggests we might be missing something. (By the standards of cosmology—which spans many orders of magnitude—30 percent is very close to 100 percent.) We need to weigh the entire universe all at once.

    B. We can do this by observing how the universe expands with time.

1. The idea is simple: As the universe expands, the matter in it exerts a mutual gravitational pull, which acts to slow the universe down.

2. Therefore, we expect to be living in a decelerating universe. The precise rate of deceleration should tell us the total amount of matter.

II. Standard candles can be used to probe the expansion history of the universe.

A. Probing the expansion history of the universe simply involves doing what Hubble did—comparing the redshifts of distant objects to their distances—but with much greater precision. For this purpose, we need standard candles that are bright enough to be visible in the very distant universe.

B. The perfect kind of candle is provided by a certain kind of exploding star called a *Type Ia supernova*.

1. When an ordinary star uses up all its nuclear fuel, it settles down to a white dwarf. White dwarfs cannot shrink smaller than a certain radius, simply because the electrons take up space (they are fermions, after all).

2. But if a nearby star dumps extra matter onto the white dwarf, a point called the *Chandrasekhar limit* is reached, at which gravity is so strong that the electron pressure can no longer resist.

a. The star collapses explosively as the electrons combine with protons to make neutrons, leading to a neutron star.

b. The outer layers of the star are blown off in the explosion, which is what we observe as a Type Ia supernova. (Other types of supernovae arise from other mechanisms.)

C. Because the Chandrasekhar limit is basically the same for every white dwarf in the universe, the brightness of the resulting explosions is basically the same, making them ideal standard candles.

1. In fact, it's better than that—although there are small variations in the brightness, these are closely correlated with variations in the time it takes the explosion to brighten and dim again.

2. Therefore, by observing a supernova over time, we can figure out precisely how bright the original explosion really was.

III. Modern techniques enable us to survey large numbers of supernovae.

   A. Supernovae are rare; only about one occurs per century in a galaxy the size of the Milky Way.

   1. Historically, that has meant that we have observed supernovae only when we've been extremely lucky.

   2. But because modern telescopes allow us to observe many hundreds of galaxies at a time, we can find supernovae almost on demand.

   B. In the 1990s, two competing groups undertook the task of measuring the deceleration of the universe using Type Ia supernovae.

   1. One group was led by Saul Perlmutter of the Lawrence Berkeley Laboratory; one was led by Brian Schmidt of Mt. Stromolo Observatory in Australia.

   2. In 1998, they both reached the same conclusion: The universe is not decelerating at all; it's actually accelerating!

   C. These measurements are very difficult to make, and it's natural to worry about whether they have been done correctly.

   1. Both groups were extremely careful to ensure that they understood various types of possible errors, such as obscuring dust or differences between supernovae in different types of galaxies.

   2. The fact that two competing groups reached the same conclusions gives us confidence that they know what they're doing.

IV. Dark energy is the best explanation for the observed acceleration of the universe.

   A. How can we explain an accelerating universe?

   1. No configuration of matter, either ordinary or dark, will do the trick.

2. We need to invent some entirely new kind of stuff—something that doesn't dilute away as the universe expands and acts as a persistent source of spacetime curvature.

3. Further, we need something that doesn't clump into galaxies and clusters but is spread smoothly throughout space.

4. That stuff, whatever it is, we call *dark energy*.

B. The good news is that dark energy completes the cosmic puzzle. If the universe is about 5 percent ordinary matter, 25 percent dark matter, and 70 percent dark energy, it fits a very wide variety of data: galactic dynamics, gravitational lensing, primordial nucleosynthesis, the cosmic microwave background, and the recent supernova surveys. It's a truly impressive model.

C. The other good news, which helped make cosmologists accept the startling news of dark energy relatively quickly, is that dark energy helps solve a number of problems at once. It reconciles the age of the universe with that of the oldest stars and implies that the total energy density is equal to (or at least very close to) the critical density.

D. The bad news is that dark energy is a mystery. It wasn't really expected and continues to be hard to explain. We need to work hard to confirm that it's really there and to provide better ideas for what it might be.

Recommended Reading:

Goldsmith, *The Runaway Universe*, chapter 7.

Kirshner, *The Extravagant Universe*, chapters 6, 8–10.

Nicolson, *The Dark Side of the Universe*, chapter 9.

Questions to Consider:

1. Even though the universe is "accelerating," the Hubble parameter is not increasing, as far as we know. Explain how this can be the case.

2. Given what we've learned so far and imagining that the universe continues to accelerate, describe the likely future of the universe.

3. Can you think of independent ways to test the dark energy hypothesis?

Lecture Fourteen—Transcript
The Accelerating Universe

We've reached that happy point in these series of lectures where we start talking about dark energy as well as dark matter. Dark energy is something different from dark matter. It is discovered in different ways; it plays a different role in the cosmic story. It comes from something different. So, the fact that we need both dark energy and dark matter to explain the observations that we see in cosmology is evidence that the dark sector of the universe, 95 percent of what the universe is made of, is interesting somehow. It's not just all the same stuff. It might be that someday we can subdivide the dark sector into more than two bits, but right now we think that dark matter plus dark energy is enough to explain everything we haven't been able to see in the universe so far.

However, the way that we get to the discovery of dark energy goes through thinking about dark matter. The thinking about dark matter goes all the way back to the 1930s when Fritz Zwicky was looking at the dynamics of clusters of galaxies. Zwicky noticed that in the Coma Cluster of galaxies, the motions of the galaxies were too fast to be explained just on the basis of the ordinary matter that you saw there. In the 1970s, Vera Rubin looked at individual galaxies and realized that they also were spinning too fast to be associated with nothing but the visible matter. So, through the '70s and '80s, people became absolutely convinced that there was something called dark matter, but the dark matter couldn't just be the ordinary matter that was hidden from us somehow. But, the question remained; exactly how much dark matter is there? Every time you look at a bigger and bigger system, you found more and more dark matter. In the 1980s, therefore, a lot of people were convinced that we would continue to find more and more dark matter. For example, you can look at individual galaxies and clusters, but how can you be sure that there isn't more stuff that is in between the galaxies and clusters? How can you be absolutely sure of that?

Furthermore, there was another reason to be skeptical, and that comes from the Friedmann equation and the notion of the *critical density* of the universe. Look again at the Friedmann equation, which relates stuff in the universe to the curvature of spacetime in the case of an expanding, smooth, homogeneous universe. On the left-hand side, you see P. The Greek letter P stands for the energy density of the universe. Now we're working again in the approximation where everything is perfectly smooth.

On the right-hand side, you see the Hubble Constant, H, which is telling us about the expansion rate of the universe, and the spatial curvature K. So, if K is zero, if space is flat, if geometry is like Euclid said it was, that's a special value for the spatial curvature. It could be positive or negative. The other terms in the Friedmann equation should not be positive or negative. They want to be positive. The energy density is something that should be positive, like positive amounts of energy, not negative amounts of energy; that's dangerous. The Hubble constant squared doesn't matter what the value of H is. It's never going to be a negative number. If you square something, you get a positive number. So, there are no special particularly interesting values of the energy density rho or the expansion rate H. But, there is a special, interesting middle value for the spatial curvature; it's zero. However, when you plug in the numbers, the energy density that we observe in the universe in matter, in clusters, and in galaxies doesn't seem to be equal to the special amount of density you would need to make the universe spatially flat. We can define the critical density as the density rho that you would need to satisfy the Friedmann equation when K equals zero, when there is no spatial curvature. We can define that density whether or not that's the density we actually have. In fact, cosmologists often define a number called omega, which is taking the actual density of the universe and dividing by the critical density. So, if the density is equal to the critical density, we say omega equals 1. In fact, with the stuff that we found in the universe, the ordinary matter and the dark matter, only about 30 percent of the critical density is there in matter in the form of clusters and galaxies. So, omega seems to be .3. That's a very strange number to have, .3. One would be a very nice number to have; that would mean the universe is spatially flat. Ten to the ten, 10 billion, or for that matter, 1 over 10 billion would be also numbers that wouldn't make you be surprised because they would just be some numbers. We can't really explain them. But, .3 makes you think you're missing something.

What .3 is telling you is that you're 30 percent of the way to being the critical density and you remember that every time you look, you find more stuff. So, throughout the 1980s, many cosmologists were convinced we would continue to find more stuff; we would eventually find enough matter in the universe to show that the density was equal to the critical density. So, .3 is close to 1, but it's not equal to 1. A lot of people just said, well, we haven't found everything yet. However, in the 1990s, that point of view became harder and harder to stick with. Technology became better; our ability to measure the energy in the universe in terms of matter became more and more convincing. Especially with things like gravitational lensing with X-ray maps of clusters of galaxies, we became convinced that we had found the matter in the universe. It wasn't adding up to 1. The idea that a cluster of galaxies is a fair sample is exactly the idea that the amount of dark matter in that cluster compared to the amount of ordinary matter is the same in that cluster as it is for the universe as a whole. If that's true, the amount of stuff that we're finding in clusters of galaxies implies that omega is only .3. It is only 30 percent of the critical density; it does not quite equal 1. So, if you're a respectable, theoretical cosmologist in the 1990s, you would have come to admit that this was true. But, I can tell you that there were very few respectable, theoretical cosmologists. The observers who actually took the data were becoming convinced that something was going on. But, the theorists were clinging to the hope that somehow omega was equal to 1.

I can remember personally giving a talk at the end of 1997, in December. I was asked to give a review talk on what are the cosmological parameters. What is the Hubble constant, what is omega, the density, and so forth? I was one of these disreputable theoretical cosmologists. I was personally convinced that omega matter must be 1; we just hadn't found it yet. But, when I sat down to look at all the papers that had recently been written, all the data that was collected, the talk I ended up giving said, "You know, something is going on. We do not live in a universe where cold, dark matter makes omega matter equal 1. Something weird is going on. Either omega is not 1, either we do not have quite the critical density, or perhaps it's not all cold, dark matter. Maybe there is a mixture of cold, dark matter and hot, dark matter, or perhaps there is something weird in the early universe that made galaxies form in a strange way. Or, maybe there is more stuff than just matter in the universe. Maybe there is the stuff, which these days we would call dark energy."

In late 1997, we were getting desperate. We had a whole bunch of things on the table for what could possibly explain the data. We didn't know which one of them was right. So, what do you want to do to resolve this? You want to weigh the universe; you want to find out how much energy density there is in space, but you want to weigh the whole universe. You don't want to just weigh a bit of it here and there because you could always be missing something in between. So, how can you weigh the entire universe all at once?

It turns out that there are two techniques that you can use. One is to actually directly measure the spatial curvature. If you measure the spatial curvature, if you measure K, you can find out whether the density that you have is only 30 percent of the critical density or whether it's 1. We'll talk about that in the next lecture. The other way is to measure the deceleration of the universe, to measure how the expansion rate of space changes as a function of time. So, this is something that you would expect to happen in ordinary cosmology. It's true that the universe is expanding; things are moving apart. But, while they are moving apart, the different particles in the galaxies, the ordinary matter and the dark matter, are pulling on all the other particles. Stuff is exerting a gravitational force. So, you expect that expansion rate to gradually slow down. If there is enough stuff, it will in fact recollapse. That would be if omega were greater than 1, if we had more than the critical density. So, if you measure precisely the rate at which the expansion of the universe was changing with time, that would tell you the total amount of stuff in the universe. The challenge is just to actually do that. It's very difficult. How do you measure the rate at which the expansion of the universe is changing? How do you measure the deceleration of the universe? You do what Hubble did, but you just do it better. Hubble found the expansion of the universe by comparing the velocity of distant galaxies to their distances. Hubble's Law, which tells us the velocity is proportional to the distance, is always going to be valid in a small region of the universe, cosmologically speaking. But, when you get out to a very, very far away galaxy, now you're looking at a galaxy that was emitting light from the distant past by the time it gets to you. You're actually probing what the universe was doing at an earlier time because light moves at only one light-year per year.

Therefore, if you measure the distances and redshifts, the apparent velocities of galaxies that are very, very far away, you can see whether or not the expansion rate has changed. You can measure the acceleration or deceleration. So, you want to do what Hubble did.

You want to use standard candles. If you have some object whose brightness is fixed—you know how bright it is—then, by seeing how bright it appears to you, you can figure out how far away it is. That's the basic idea of a standard candle. For Hubble, the standard candles he used were Cepheid variable stars, pulsating stars for which you could figure out from the period of pulsation how intrinsically bright the star was. The problem is that Cepheid variables are not that bright. They are the brightness of ordinary stars. You can't pick out individual Cepheid variables that are in very, very distant galaxies. Instead, you need a much brighter standard candle. Eventually, what you appeal to are supernovae, exploding stars that are incredibly bright. Here's an image of one of the most beautiful supernovae you'll ever see. This is supernova 1994D and you can see that there is a galaxy; the supernova is in the bottom left. That is a star in that galaxy that is not a nearby star in our galaxy. The brightness of that supernova is comparable to the entire brightness of the galaxy behind it. It's billions of times the brightness of an ordinary star. That's the good news. The bad news is that they are rare. You don't see supernovae all the time. You can't predict them. In a galaxy the size of the Milky Way, you're only going to get about one supernova per century. The other problem is that supernovae are not standard candles all by themselves. They are not all the same brightness. There are different kinds of supernovae. In fact, you remember that we mentioned when we were talking about MACHOs, how you create neutron stars, that there is a type of supernova called a core collapsed supernova. You have a bright, heavy star burning nuclear fuel. The nuclear fuel burns out and the core of the star just collapses, it explodes off the outer layers and that is a Type II supernova. Clearly, for different masses of stars, when they collapse, their brightness is going to be different.

Type II supernovae are not standard candles, but from the name, Type II supernovae, you might guess there is something called the Type I supernova. In fact, there are various different kinds of Type I supernovae, and there is a particular type called Type Ia that can be used as a standard candle. A *Type Ia supernova* is a very different object than a Type II. A Type Ia comes from a white dwarf star, so a white dwarf is what you get when a medium mass star gives out its nuclear fuel. It just settles down to be a white dwarf. But, imagine that you're lucky and you have a white dwarf star that has a companion. There is another star next to it, which in the course of its evolution, grows and the white dwarf begins to accrete some of the mass from its companion.

Here's a picture of an artist's conception of a white dwarf getting mass away from a nearby star, and so the mass of the white dwarf gradually grows and grows and grows. But, there is a limit. White dwarfs cannot be arbitrarily massive. Eventually, the gravitational field will become so strong that the white dwarf collapses. This limit is called the *Chandrasekhar limit*, and it's the same for every white dwarf everywhere in the universe. So, you can see you have a hint that something, in fact, is a standard candle. The place where the white dwarf collapses and the outer layers are blown off forms a Type Ia supernova. It would not be surprising if every such event was more or less the same brightness, and it's true. They are plausibly standard candles. Type Ia supernovae could all be approximately the same brightness.

But, there are various problems associated with the idea of using Type Ia supernovae to measure the acceleration or deceleration of the universe. First, Type Ia supernovae are not precisely standard candles. By looking at nearby supernovae, people noticed that Type Ia's differed in brightness by about 15 percent. Fifteen percent doesn't sound like that much, but when you're trying to look for a very, very subtle change in the expansion rate of the universe, every 15 percent counts. The real breakthrough in this field came when Mark Phillips in the late 1990s realized that, just like Cepheid variables, Type Ia supernovae have a period luminosity relationship. The supernova doesn't pulsate, but it does go up in brightness and then go down. What Phillips realized is that the time it takes to decline in brightness told you what the maximum brightness was. The Type Ia supernovae that are the brightest are those that take the longest to decline. So, if you measured not only the maximum brightness of the supernova, but also how it evolved, how the brightness declined as a function of time, you could really pin down that overall brightness to better than 5 percent. Then, you have something that is a good enough standard candle to measure the deceleration of the universe.

The other things are more like worries. One is how do you know, when you observe a supernova very far away, that they were just as standard in the early universe as they are today? That is something we're going to have to deal with by taking a lot of data and trying to figure out what the physics behind these objects are. But, more importantly, how do you find them? You could look at a galaxy and stare at that galaxy for a hundred years, and then you have a 50 percent chance of finding one supernova. No one is going to give you telescope time to do that. You need to come up with a better technique. The thing is that only in the 1990s did astronomical technology evolve to the place where we could find a whole bunch of supernovae all at once. So, people developed techniques using large CCD cameras. A CCD is a charged coupled device, which allowed you to take an image of a fairly wide swath of the sky. You want to take an image that is deep enough to get lots of galaxies, but wide enough that you can get a whole bunch of them so you get galaxies of different redshifts in great numbers. What you do is you notice that the rise time of a supernova, the time it takes to go from being dim to being very bright, is a couple of weeks. That turns out to be a very convenient time. You take an image of some region of the sky with literally thousands of galaxies in it and then you come back again a couple of weeks later to take another image. In fact, you can do the first image at new moon where the sky is the darkest, and then you take the next image at the next new moon. It works out perfectly. Then, you want to compare these two images to look for one of the galaxies getting a tiny bit brighter.

So, this picture is a little bit of a fake because this picture is a better-than-average view of what such a supernova discovery looks like. That's because this picture is from the Hubble space telescope, not from the ground-based telescope where most of this work is done. But, you get a feeling for what's going on. You see an image on the left of the deep field taken by the Hubble space telescope and another image of the same region of the sky taken several years later. You can tell there is a supernova in the second image because there is an arrow pointing to it. That's the nice thing about these images. There are always arrows pointing to where you care about. In this case, the arrow points to a fairly dim red dot that you can zoom in on and find it is, in fact, a supernova all by itself. This is a very distant supernova. This is an especially good example, but using this kind of technology, you can find supernovae by the dozens.

This project of finding a whole bunch of supernovae, Type Ia's, and using them to measure what the universe was doing at earlier times was undertaken in the 1990s by two different groups. One group was centered at Lawrence Berkley Laboratory and was led by Saul Perlmutter. The other group was scattered around the globe; it was not really centered anywhere, but the leader of the group was an Australian astronomer named Brian Schmidt. Here's a picture of Brian Schmidt on the left and Saul Perlmutter on the right, arguing over whose universe is accelerating or decelerating faster. It was a mostly friendly rivalry between the two groups, Perlmutter's on one side, Schmidt's on the other. Schmidt's group involved a lot of people. Adam Riess, who is now at the Space Telescope Science Institute, was the lead author of the most important paper. Bob Kirshner, who was at Harvard, was both Brian Schmidt's advisor and Adam Riess's advisor and sort of the intellectual godfather of the team. Alex Filippenko was a prominent member of the team who's most famous for giving Teaching Company lectures on modern astronomy that I encourage you to have a look at.

It was important that two groups were doing it because if only one group did it, nobody would believe them. If two groups do it and they get the same result, then people are ready to think that something is going on that's on the right track. Indeed, it was in 1998, only a couple of months after I gave my personal talk in 1997 when I said something is going on, that the supernova groups in 1998 announced that something was going on, mainly that the universe was not decelerating at all. It was accelerating. It was expanding faster and faster. The correct image of the universe is one in which galaxies have a velocity that is increasing as a function of time. This came as a great surprise. Most people did not expect we would live in an accelerating universe. The Schmidt group, in fact, had a subtitle, "Searching for the Deceleration of the Universe." What they found was the acceleration of the universe. So, you had a very strange situation where, on the one hand, the result was a complete surprise. On the other hand, it made perfect sense. The reason that people were willing to believe this result, besides the fact that two very good groups got the same result, was that it made things snap together. It answered a lot of questions all at once, as we will explain.

However, the fact that the universe is accelerating is a physical challenge; it's not what you expected. There is an intuitive argument that, as the universe expands, it should be decelerating because particles are pulling on each other. If particles should be pulling on each other, how in the world do you explain an apparently observed phenomenon that the universe is accelerating rather than decelerating? The answer is you need something besides particles.

You need to invent a new kind of stuff.

If the Friedmann equation is correct and the universe has nothing in it but matter and radiation, it doesn't matter what kind of matter and radiation you have. The universe will necessarily be decelerating. So, either the Friedmann equation is not correct, according to these data—and we will explore that possibility later—or there is something in the universe that is neither matter nor radiation. We call that stuff dark energy.

What does *dark energy* mean? We use the word dark energy; it sounds a little bit mysterious. But, I want to emphasize that even though there is a lot we don't know about dark energy, it's not just a place holder for something going on we don't understand. The dark energy really does have some properties that are absolutely definite and need to be part of any theory of what the dark energy is. First, the dark energy is smoothly distributed through space. It is more or less the same amount of dark energy here in this room as it is somewhere in between the galaxies. At the very least, dark energy does not clump noticeably in the presence of a gravitational field. The reason why we know that is that if it did, we would have noticed the dark energy before when we measured the energy density of galaxies and clusters. By gravitational lensing and by other means, we would have seen that there is more energy here than can be explained by the matter and that would be the dark energy. But, we don't see that. The dark energy is the same amount inside a cluster as outside a cluster. That's why we didn't see it when we looked with gravitational lensing and dynamical means. So, you might guess that the dark energy could be some form of radiation, that it could be something that is moving very quickly. If something is moving very quickly like photons or neutrinos, it would not cluster into galaxies and clusters, so that would be smoothly distributed just like the dark energy. But, the second important property of dark energy is that it is persistent. The energy density, the number of ergs per cubic centimeter of the dark energy, doesn't change as the universe expands. That's the opposite of what radiation does. Radiation loses energy rapidly as the universe expands. For dark energy, you need something that doesn't go away. It is the fact that dark energy is persistent that explains the acceleration of the universe. That's why we're convinced that the dark energy is really something different. It's not a kind of particle that is pulling on other particles. The dark energy is a kind of field, the kind of substance, the kind of fluid that fills space. Its energy density doesn't go away as the universe expands.

But, this is asking a lot. This is going around saying, OK, we've discovered some stuff that is completely unlike ordinary matter, dark matter, or radiation. Nevertheless, over the course of 1998, people began to buy this story fairly quickly. Why are astronomers, who are by nature quite skeptical people, willing to believe this remarkable claim? The real point is that it made everything suddenly make sense. Like I said, in 1997, things were becoming difficult to understand. We had a prejudice that omega matter, the density in ordinary stuff plus dark matter, should be 1, that we should have the critical density. But, the observations simply weren't consistent with that. Plus, there were other problems that made the universe in which we believe not quite make sense when compared with the data that we had. One problem was the age problem. Given the amount of stuff in the universe and given the Friedmann equation, you can calculate how old the universe should be. It's an absolute requirement that the universe should be older than the stuff inside. When people calculated the age of the universe, when they compared it to the ages of the oldest stars, they were often getting an answer that the stars were older than the universe. That didn't quite make sense. There were large error bars on that; it was not a very hard and fast conclusion, but it still made you worry that you were missing something.

In an accelerating universe, the universe today is older for any given value of the Hubble constant than it would be in a decelerating universe, so it made the age problem go away just like that. Another problem was large-scale structure in the universe. If you had a universe with nothing but matter in it and the matter was the critical density, you form a lot of structure. There is a lot of matter around, it clumps very easily, and you have more structure in that hypothetical universe than we seem to be finding in the universe which we observe. You could explain this by saying that we don't have that much matter, that we don't have the critical density. But, another way to explain it is to imagine there is something that doesn't clump.

There is some dark energy that is smoothly distributed that doesn't contribute to the growth of large-scale structure.

Finally, there is this business about the critical density. Like I said, there was a prejudice on the part of theorists that the critical density was the nicest value for the total density to have. They were therefore hoping that observers would continue to find more matter, even though the observers were telling them, no, we haven't found any more. It turns out that once you find that the universe is accelerating and you invoke the presence of dark energy as an explanation for this acceleration, you ask, well, how much energy do you need in the dark energy? The answer is about 70 percent of the critical density. In other words, the really nice thing about the dark energy was that it provided exactly enough energy to make the total energy density of the universe equal the critical density. So, it wasn't just that we had found a new element of the universe, it's that we had found what was apparently a complete picture, a complete inventory of the universe.

This is where this pie chart that we began our lectures with comes from. We have what is now called a *concordance cosmology*, a view of the universe in which 5 percent of the stuff in the universe is ordinary matter, 25 percent is dark matter, and 70 percent is dark energy. That simple set of ingredients is enough to make the universe flat, to be the critical density, to get the age right, to correctly explain large-scale structure, the acceleration of the universe, the cosmic microwave background, and get the matter density right in galaxies and clusters. That is a lot of observations that come from a small amount of assumptions. That's why people were so quick to jump on to the concordance cosmology bandwagon. It also tells us something about our place in the universe. Not only are we not, like Aristotle would have it, sitting at the center of the cosmos; we are not even made of the same stuff as the cosmos. We are only five percent of the universe. The kinds of things we're made of are only five percent of the energy density of the stuff of the universe. This is a big deal. It is a sufficiently big deal that it was recognized by *Science Magazine* in 1998 as the breakthrough of the year. They invented a nice little picture of Albert Einstein blowing the universe bubbles with his pipe to illustrate the fact that this dark energy was in fact something that Einstein himself had contemplated, as we will talk about in later lectures.

Of course, even though it makes everything fit together, the dark energy and its role in the concordance cosmology are still dramatic claims. We would certainly not accept the evidence of the supernovae data as sufficient to believe in this dramatic thing. We want to check things. For one thing, we want to check that the supernovae are telling us the right thing. For example, the statement that the supernovae indicate the universe is accelerating is just the statement that very distant supernovae are dimmer than we would have thought. So, you can invent much more mundane ways to make supernovae dimmer. For example, maybe there is a cloud of dust in between us and the supernova that is actually just scattering some of the light. That is absolutely the kind of thing that the supernova groups took very seriously. The point is that when dust scatters light, it doesn't scatter every wavelength equally. It scatters more blue light than red light. So, the light from the supernova would get reddened if it passed through dust. One of the things that the supernova groups did was to check very carefully for reddening and for other alterations of the spectra of the supernova they were looking at, and they didn't find any. They also checked that the behavior of the supernova they observed was the same, whether they came from small little galaxies or big galaxies, whether they were in clusters or outside clusters. There was no environmental effect that was leading to the supernova being different one place or another. The real killer check, however, on this picture of the universe would be to make a prediction using it and then to go measure that prediction. So, there is a prediction made from this concordance cosmology. If 5 percent of the universe is ordinary matter, 25 percent is dark, and 70 percent is dark energy, and it all adds up to the critical density, then the spatial curvature of the universe should be zero. The universe should be spatially flat.

So far, we haven't talked about any observational check on that, so the next lecture will take us through how we know whether or not the universe is spatially flat. The answer is, yes, it is spatially flat. In other words, even without the supernova data, something is 70 percent of the energy density of the universe, and that something is the dark energy. That's good news. It's good news that we understand something new about the universe: 70 percent of it is dark energy. The bad news is we don't know what that stuff is in some deep level. So, in addition to making sure that we're on the right track that there is dark energy, theoretical physicists are now faced with the task of explaining what the dark energy is, where it came from, what it might be going into, and how it can interact with other stuff. There is a simplest guess, which is Einstein's idea of the cosmological constant. We will take that guess seriously. We will also look at alternative explanations to see which one fits the best.

Lecture Fifteen
The Geometry of Space

Scope:

Although the concordance cosmology fits the supernova data, it has nevertheless been necessary to perform independent tests. Fortunately, one test is provided by the cosmic microwave background. Fluctuations in the CMB have a characteristic size on the sky, one whose angular extent is related directly to the curvature of space and, thus, to the total amount of energy. Soon after the supernova results were reported, measurements of the CMB became precise enough to determine the total energy of the universe, with a result that fit the prediction of the concordance model precisely. The need for dark energy had been confirmed by a completely independent technique.

Outline

I. Observations imply that our universe is spatially flat.

    A. In Lecture Five, we discussed how the Friedmann equation of general relativity related the expansion rate, energy density, and spatial curvature of the universe. We've now measured the total energy density (matter plus dark energy), as well as the expansion rate (the Hubble constant).

        1. If these observations are correct, there should be a testable prediction for the spatial curvature.

        2. When we run the numbers, we find an interesting answer: Given the amount of ordinary matter, dark matter, and dark energy implied by the data, space should be very close to flat (if not exactly so).

    B. A flat universe, rather than one that is noticeably curved, has always been the favorite possibility among theoretical cosmologists.

        1. Not only is it an aesthetically pleasing middle point between positive and negative curvature, but it is a strong prediction of the inflationary universe scenario (to be discussed in Lecture Twenty).

2. In addition, we have the *flatness problem*: In a matter/radiation-dominated universe, the curvature grows with respect to the energy density as the universe expands.

3. Of course, dark energy was never a favorite possibility among theoretical cosmologists; thus, their preferences should be taken with a grain of salt.

II. We can measure the curvature using the cosmic microwave background.

A. To measure the total curvature of space, we should reproduce (on a much larger scale) the kind of experiment performed by Gauss, as described in Lecture Five: Add up the interior angles of a really big triangle and see if they sum to 180 degrees.

B. Fortunately, the universe provides us with just such a triangle. One vertex is us here on Earth, and the other two are located on the surface of recombination, as seen in the CMB.

1. Remember that the CMB temperature fluctuations are largest for those regions that have just had time to collapse but not to re-expand.

2. Given that recombination happened around 380,000 years after the Big Bang, we know very well how big those regions are: They are 380,000 light-years across (up to some small numerical factors that don't concern us here)!

3. We also know how far away the CMB photons come from, giving us all the information we need about our triangle.

C. In a flat universe, the largest fluctuations in the CMB should be about 1 degree across. Positive curvature would lead to a larger angular size, and negative curvature would lead to a smaller size.

1. Observations from the Wilkinson Microwave Anisotropy Probe (WMAP) and elsewhere are perfectly clear: The primary anisotropies are, indeed, 1 degree across.

2. Space is flat, just as the concordance cosmology dominated by dark energy says it should be.

III. The CMB observations of the geometry of space provide a stunning confirmation of the dark energy picture. Although it is a bold and startling idea, it (or something even more surprising) seems forced on us by the data. But dark energy is quite a puzzle. Why does it have the magnitude it does? Why is the amount of dark energy similar to the amount of matter? And, ultimately, what is dark energy?

Recommended Reading:

Goldsmith, *The Runaway Universe*, chapter 11.

Kirshner, *The Extravagant Universe*, chapters 6, 8–10.

Lemonick, *Echo of the Big Bang*, chapter 9.

Questions to Consider:

1. By observing the curvature of the universe, we claim to have measured the *total* energy density, without anything missing. Is there any way that presumption could be mistaken?

2. Besides using the CMB, can you think of any other ways to measure the universe's spatial curvature?

Lecture Fifteen—Transcript
The Geometry of Space

We're in sort of a good news/bad news situation right now with where we are with the concordance cosmology. The good news is we have a model that fits all the data: 5 percent of the universe is ordinary matter, 25 percent is dark matter, and 70 percent is dark energy. Since 1998 when we first found evidence that the universe is accelerating, this idea has been tested in many, many different ways. It keeps coming out passing the tests. It seems as if the universe is accelerating just as the first supernova observations indication.

The bad news is we don't understand it. The dark energy part in particular is very exotic and outside our ordinary experience. In fact, the dark matter part is also very exotic. Ten years ago, I would be giving a set of lectures just on dark matter, and that would also seem very exciting and exotic and interesting. These days, the dark matter seems almost prosaic compared to the dark energy, which is something truly different. Clearly, we want to do the best we can to test this idea, to figure out whether or not this hypothesis, 70 percent of the universe is some smoothly distributed, persistent form of energy density, is in fact correct. We want to use that hypothesis to make predictions, and then go out there and test those predictions.

The most obvious test that we can think of is the one that I mentioned in the last lecture, the geometry of space. Here is, once again, the Friedmann equation, the equation derived from Einstein's theory of general relativity that governs the relationship between energy density and the curvature of space in an expanding universe. What we're saying here is that the energy density is given by the sum of a contribution from the expansion and a contribution from the curvature of space itself. It turns out that the right amount of dark energy you need to make the universe accelerate is also the right amount of dark energy you need to make the universe spatially flat. In other words, to satisfy the Friedmann equation with K, the spatial curvature term, being zero. That is a testable prediction. We can try to imagine measuring the curvature of space itself.

In this lecture, we're going to go out and actually measure the curvature of space on very, very large scales. In other words, what we're doing is the same kind of thing we were doing when we were using gravitational lensing to measure the weight of a galaxy or a cluster of galaxies. What we were doing then was letting light rays go by a cluster or some other massive object and we were mapping out the local geometry of space near that gravitating object. Since Einstein tells us that space and time are warped and bent by the presence of matter, mapping out the geometry of space and time is a way to tell how much stuff there is. So, we can do that very straightforwardly with individual clusters of galaxies. Now we're going to do that with the whole universe. We're going to look at how light propagates through space to measure the geometry of all of space all at once. That's how we can be sure that we're not missing anything. When you're looking at individual clusters, it's always possible in principle that there is something outside. When you look at the whole universe, you're guaranteed to find everything that there is.

Here are the possibilities for what the curvature of space could be. Remember what this means. This means that since the universe is uniform everywhere, there is a certain, single number which tells us how much space is curved. That number could be positive, it could be zero, or it could be negative. That number is the curvature of space. It's related to omega, the density parameter. Omega is telling us what the total density of the universe is divided by the critical density in the sense that when omega is one, the universe is spatially flat. When omega is greater that one, it is positively curved like a sphere. When omega is less than one, space is negatively curved like a saddle. You can think about the directions of curvature of space bending in different directions.

The curvature of space is something like many other things in cosmology; it is hard to visualize the reality of it. Whenever you are drawing pictures, representations of the curvature of space, the best you can do is draw a two-dimensional version. The actual space in which we live is three-dimensional. As far as we know, the three-dimensional space in which we live is not embedded in any bigger space. It is all there is. So, when we're talking about the geometry of space, we're talking about the intrinsic geometry of space. We're not talking about how it looks to an outside observer; we're talking about things you can do when you are inside space to measure its geometry.

When we say the geometry is flat, we mean that the kind of geometry that Euclid invented thousands of years ago is the right kind to describe space on very large scales. Euclid looked at tabletops as a paradigm for where to do geometry. He would draw right angles and triangles, and come up with different laws of how geometric shapes fit together. We want the three-dimensional version of that, so we want to draw triangles and straight lines and circles within a big three-dimensional space, and look at their intrinsic properties. So, they are different intrinsic properties that are telling you how space is possibly curved.

The most obvious one is the angles inside a triangle. If you draw a triangle on a tabletop or a triangle just in a three-dimensional space that is flat, no matter how you draw that triangle, no matter how it's oriented or no matter how big it is, when you add up the angles of those three angles inside, you will always get 180 degrees. The way to think about the statement "space is flat" is to translate it into the statement "every single triangle I can draw anywhere in the universe has angles inside that add up to 180 degrees." If space is positively curved, a similar statement would hold true about triangles, but it would say that the angles would always add up to more than 180 degrees. We can visualize that by imagining a sphere, drawing a triangle on the surface of a two-dimensional sphere, it would be the case that the angles inside add up to greater than 180. So, we have the three-dimensional version of that. We're imagining that the space itself in which we live is a three-dimensional version of the sphere. Just like on a regular sphere, if you start at one point and walk around, you will eventually come back to where you left. In a three-dimensional sphere, in a universe with positive spatial curvature, if you walk off in one direction, you will eventually come back to where you left. If will take you billions of years, but you would eventually get there.

Negative curvature, then, means that you once again draw big triangles. In a negatively curved space, every triangle you can draw has angles inside that add up to less than 180 degrees. That's the kind of thing you can do. It's not the only thing you can do. A famous postulate of ordinary Euclidian geometry is the parallel postulate. It says if you start with two straight lines and let them go, two parallel lines, if they are initially parallel, they will always remain exactly parallel. That is to say, if they are initially having the property that the distance between the two lines is a constant and not changing as you go past the lines, that will always be true no matter how far you go. That is a postulate of Euclidian geometry. It's not a necessarily true fact about the world.

If you lived in a positively curved space, two lines that were initially parallel would eventually come together. That makes perfect sense thinking about a sphere. You can draw two lines as parallel as you want. If you trace them down, they will eventually hit each other somewhere. In a negatively curved space, meanwhile, you start with two parallel lines. You follow them down, and guess what? They are going to peel off. They are going to become more and more far apart. That is a thing that you can imagine empirically doing in space that would reveal whether space was positively curved, negatively curved, or flat.

So, we've briefly touched upon something called the *flatness problem* in actual cosmology. The flatness problem has sort of an informal version and a formal version. The informal version of the flatness problem is the universe seems to be close to spatially flat. When you calculate the actual density of matter in the universe now, putting aside dark energy for a bit, if you knew only about the matter, you would say that we've reached 30 percent of the critical density necessary to make the universe flat. As these things go, 30 percent is awfully close to 100 percent. It makes you think that we're almost there. So, there is this feeling that you're just missing something. You should be exactly flat if you knew what everything was. That's the sort of informal version of the flatness problem.

But, there is a more formal version of the flatness problem, a statement that is a little bit more quantitative that drives home exactly how surprising it is that the universe is close to being spatially flat without exactly being there. Basically, there is a statement that the universe doesn't want to be flat. If the universe is a little bit non-flat, if there is a little bit of curvature to space, the amount of curvature becomes more and more important as the universe grows. So, if you live in an old universe, the universe should be very curved if there was any curvature at all. You can see this roughly from the Friedmann equation, once again. The Friedmann equation is relating the energy density to the expansion rate and the spatial curvature, so you have three terms in that equation. But, remember that for the energy density, we have very well defined rules about how the energy density changes as the universe expands. For ordinary matter or for dark matter, for any particles that are slowly moving compared to the speed of light, as the universe grows, the density goes down and the volume goes up. So, the energy density goes down exactly like the number density goes down. The particles just become more dilute.

For radiation, the energy density goes down even more quickly. Because the number density goes down, space becomes bigger. But, also, every single particle of radiation is losing energy. So, we have very well defined rules that tell you how the energy density changes as the universe expands. There is also a well-defined rule that tells you how the spatial curvature changes as the universe expands. You can imagine blowing up a balloon. The balloon, when it's very small, looks very curved. As it gets bigger and bigger, every little cubic inch in the balloon makes it look flatter and flatter. In the same way, curvature dilutes away as the universe expands.

However, when you plug in the numbers, curvature dilutes away more slowly than the energy density in matter or radiation. To be technical, the curvature goes like one over the scale factor squared. The energy density in matter goes like one of the scale factor cubed and the energy density in radiation goes like one over the scale factor to the fourth power. So, as the universe gets bigger and bigger, as the scale factor grows, the energy densities in matter and radiation fall off more rapidly than the contribution to the spatial curvature. So, imagine that there is some non-zero spatial curvature in the very early universe, that the term K in the Friedmann equation is not exactly zero at early times. Even if it's fairly close to zero, the relative importance of that term compared to the importance of the energy density in matter or radiation grows. We live today in a very old universe. The universe is 14 billion years old. It's had 14 billion years for the spatial curvature to overtake and overwhelm the energy density of matter and radiation. But, it hasn't. That is the technical statement of the flatness problem. In order to get a 30 percent of the critical density universe today in just matter and radiation, the difference between being absolutely flat and being curved in the early universe had to be incredibly infinitesible, had to be very finely tuned to just the right tiny, tiny, tiny amount, so that today it would be comparable to the energy density in matter and radiation.

This seems like an unlikely situation. That is the flatness problem. So, 30 percent of the critical density, the amount that we found in ordinary matter, is a weird number to have. That's why before we found the dark energy, theoretical physicists who were very convinced by this were thinking that omega matter, the contribution of the total density of the universe just from ordinary matter plus dark matter, had to be one. They believed that the universe had to have the critical density. Since it was so close, it would make no sense to not quite go all the way. They believed in a flat universe. But, at the end of the day, you can believe whatever you want. It's not going to make you any money. You've got to go out there and look.

So, how do you actually measure the spatial geometry of the universe? Ideally, you want to do what Carl Gauss did when he was inventing the concepts of non-Euclidian geometry. He actually made a big triangle and he measured the angles inside. So, we want to make a big triangle in the universe. Of course, we can't travel to distant galaxies stringing lasers or a piece of wire from one place to another. We have to take celestial objects as they are given to us and use them to construct a big triangle, add up the angles inside. One way to do this is if we had a standard ruler, not a standard candle. A standard candle is something whose brightness we know and so the further away it is, the dimmer it will look. A standard ruler is something whose size we know. So, the further away it is, the smaller it looks. So, a standard ruler is just as good a way to measure distances to objects as a standard candle is. The reason why you don't hear as much about standard rulers is because there just aren't as many of them. There aren't that many objects in cosmology or in astrophysics whose size, in miles across, you actually know ahead of time. Galaxies, stars, and different astrophysical objects appear in different sizes.

However, imagine that you not only had a standard ruler, but imagine that you knew how far away it was. Imagine you had some object whose size you knew and whose distance you knew. Then, you would think, I must be able to figure out exactly how big it would look. That would be true, assuming that you knew the geometry of space. The angular size that that ruler will take up, if you know how big it is and you know how far away it is, is telling you the geometry of space. If it has a certain angular size, one degree across in a flat universe, then in a positively curved universe it will look bigger than one degree across. That's because the angles that you're subtending from the light rays that come from you to the object or vice versa are pulled together by the positive curvature. So, they can start out bigger and they will end up at the different angles of your standard ruler.

Similarly, if you have a negative curved universe, that ruler which would have looked one degree across is now going to look smaller than one degree across. However, this is of course asking a lot. We're asking not only that we have some object whose size we know, but we want to put that object in some place and we know exactly how far away it is. How lucky do we have to be to have some object that has a fixed size and a known distance? Well, we got lucky. The universe provides us with exactly that in the form of the temperature fluctuations in the cosmic microwave background.

We already talked about the cosmic microwave background, the leftover radiation from the Big Bang. The cosmic microwave background is a snapshot of what the universe looked like about 400,000 years after the Big Bang. In times earlier than that, the universe was so hot that individual atoms were ionized and the universe was opaque. Light could not travel very far before bumping into an electron. After 400,000 years, the universe had cooled down enough that electrons and nuclei had gotten together. The universe became transparent and light traveled unimpeded through the universe. So, we see what the universe is looking like about 400,000 years after the Big Bang.

The universe at that early time was much smoother than it is today. Today, the universe is smooth on large scales, but on the small scales, we see individual planets, stars, galaxies, and clusters. At very, very early times, the universe was smooth on essentially all scales. It's the tiny deviations from perfect smoothness at that early time that grew under the force of gravity into stars and galaxies and clusters. So, if you look at the cosmic microwave background, observing not only that it exists but also delicately measuring the temperature of the cosmic microwave background at different points in the sky, you're measuring the imprint for those primordial fluctuations in density.

Here is the classic picture of the cosmic microwave background from the WMAP satellite. It is an all sky picture. It is the whole sphere that we look at when we look at the sky projected onto an ellipse in this image. What you're seeing here is just the tiny fluctuations in temperature. These fluctuations from place to place are only 1 part in 100,000. The blue spots are a little bit cooler; the yellow and red spots are a little bit hotter. This is telling us where the *density fluctuations* were located on a sphere 400,000 years after the Big Bang.

So, we can learn a lot from these density fluctuations. We don't know any theory that predicts a precise place for where they should be, and we don't ever expect to have such a theory, but we do know the statistical properties that they should have. We already, in an earlier lecture, talked about the fact that the properties of these splotches on the microwave background—hot spots and cold spots—provide evidence for dark matter. That's because what you have in the primordial universe is a hot plasma that is experiencing acoustic oscillations. You have an ionized plasma, but in some regions it's a little bit more dense and in some regions it's a little bit less. Under the force of gravity, a region is going to collapse under its own gravitational field and heat up. That will increase the temperature fluctuation in that region. It's now hotter than it used to be.

The dark matter, of course, continues to fall in, but the ordinary matter bounces because of pressure and it becomes less hot than it used to be. That's also a fluctuation, but now it's a lower temperature than a higher temperature. It goes back and forth exactly like that. Let's ask the question, what size would you expect the largest amount of fluctuation to be? The point is if a region that is going to collapse is very, very big, it doesn't have time to collapse in any appreciable way. Its temperature fluctuation is going to be whatever it was stuck with in the very, very early universe. If it's too small, the region collapses and expands and collapses and expands. It bounces back and forth and gets damped. So, eventually, it settles down into a configuration that is not very fluctuated. It is not very different from the surrounding plasma.

However, there is one particular length for which a region has time to collapse and heat up, become fairly substantial in terms of the difference that it has in temperature, but doesn't have time to bounce back in. That will correspond to a physical size in light-years of the age of the universe in years at the moment when the microwave background is formed. In other words, since the microwave background is a snapshot of 400,000 years after the Big Bang, regions that are 400,000 light-years across will have the greatest amount of fluctuation in their temperature. So, we have a prediction. We have a prediction for, at that moment, in the cosmic microwave background, we know how big the most noticeable hot spots and cold spots should be. They should be 400,000 light-years across.

Furthermore, we know how far away the cosmic microwave background is. Using the Friedmann equation, using some theory for what the universe is made of, which we have, we can predict the distance from here, us today, to there, 400,000 years after the Big Bang. In other words, we have all the ingredients for making a big triangle. We have a standard ruler on the microwave sky. It's the hot and cold spots with the largest amplitude and they are predicted to be one degree across if the universe is spatially flat. In other words, if the universe is spatially curved, positively or negatively, that's like taking the map that you would predict in a flat universe and either magnifying it or shrinking it. So, these are the data that you have from the WMAP. You can compare them to the prediction. What is the answer? The answer is that the universe is spatially flat. This is, in fact, not first found by WMAP. There were other experiments that found this.

The boomerang experiment from Antarctica was actually the first to make a precision measurement of the size of the most dominant fluctuations in the cosmic microwave background. They all knew ahead of time what they were looking for. If it was one degree across, that would be evidence that the universe was spatially flat. This was in the year 2000, soon after dark energy had been discovered, when people were still very excited although skeptical. They wanted to know, was this picture hanging together? So, the boomerang experiment and other experiments following up on it looked for fluctuation to cosmic microwave background one degree across. That's exactly what they found.

In other words, the microwave background is telling us that the universe is spatially flat, that the total density of stuff in the universe is the critical density. It fits perfectly with the concordance cosmology of 5 percent ordinary matter, 70 percent dark energy, and 25 percent dark matter. So exciting was this that in 2003, when WMAP came along, *Science Magazine* declared it to be the breakthrough of the year, even though what they really did was just say, we didn't make a mistake in 1998. In 1998, the acceleration of the universe was the breakthrough of the year; that suggested the concordance cosmology. In 2003, WMAP came along and said, yes, that's right. This crazy universe with 5 percent, 25 percent, and 70 percent is really the universe that we live in. This was sufficiently surprising that, again, that was the breakthrough of the year.

One thing to take very seriously is the fact that the microwave background tells us the total energy density of the universe. So, if you just take evidence from the cosmic microwave background that the total energy density is the critical density, and you add that to the evidence from clusters of galaxies, ordinary galaxies, and gravitational lensing, that matter only adds up to 30 percent of the critical density, then you don't need to mention the word supernova to be convinced that there is something called dark energy that is 70 percent of the universe. It's just 100 percent minus 30 percent is 70 percent. In other words, whether or not we have a correct theoretical understanding of the physics of Type Ia supernovae, whether or not the two supernova groups did a good job in explaining their error bars and collecting data that was reliable, you're still forced to the conclusion that dark energy exists.

The constraints that we have right now over-determine the kind of universe in which we live. You can't wriggle out of the conclusion that there is a lot of dark energy just by being skeptical of the supernova results. So, the place we are is that we're stuck with a universe in which 70 percent of the energy density is dark energy. We now must face up to the same kind of problem that we had with dark matter. Given that there is this stuff, what could it be? Let me just foreshadow a little bit by saying what the simplest possible candidate for dark energy would be, and that is something called vacuum energy. Remember that the two important properties we know dark energy has are that it is spatially uniform, more or less the same in every location in space, and it's persistent in time. The density, the amount of energy per cubic centimeter, isn't changing very much for the dark energy as the universe expands. So, the simplest idea for something that is more or less smooth in space and more or less constant in time is something that is exactly constant throughout both space and time.

In other words, it's a form of energy density that is inherent in spacetime itself. Just the statement that every little cubic centimeter of space contains energy, whether there is any stuff in that cubic centimeter or not. So, here is the closest I could come to an artist's representation of what this idea, vacuum energy, really looks like. You imagine you take a little cube of completely empty space and you ask yourself the question, how much energy is there in that cubic centimeter of empty space? According to general relativity, there is no reason why the answer has to be zero. The energy density of empty space is a constant of nature. It could be negative, it could be positive, or it could be zero. The number, whatever it is, is the vacuum energy. So, the hypothesis that that energy is the correct value to explain 70 percent of the critical density fits with the hypothesis that that is the dark energy.

The dark energy is, in fact, the energy density of empty space, the vacuum energy. This is the same thing that Einstein was talking about when he first invented what he called the cosmological constant. Einstein added this term to his equation, as we will talk about later, because he was not able to explain a static universe within his theory. We now know the universe is not static, so Einstein said that was a great mistake of his, but now we're bringing back the cosmological constant that's exactly equivalent to this idea of vacuum energy.

That's a very good idea. We don't need to complicate the idea. The idea that the dark energy is vacuum energy inherent in empty space, plus our ideas about dark matter and ordinary matter, are enough to fit the data. Why then do we even contemplate other possibilities? Well, there is one point, which is that we were just surprised. We were very surprised with the existence of dark energy itself. We had a preference. We had a prejudice ahead of time that matter made up the critical density and we were wrong. So, our prejudices about what ideas are simple and make sense shouldn't be taken too seriously. We're keeping an open mind about this possibility.

In fact, just like the flatness problem, there are sort of naturalness and fine-tuning problems associated with the concept of a vacuum energy. One is called the *cosmological constant problem*, which is just the statement that once you admit there can be vacuum energy, you can start asking how big should it be. The answer, as we will see, is that it should be much bigger than what we would observe. Once you admit that there can be any energy density in empty space, the surprise is that it's so tiny. The other problem is called the coincidence scandal. The coincidence scandal is an exact analog to the flatness problem. Remember, the flatness problem said, look, the

©2007 The Teaching Company.

energy density of stuff decays away very quickly. The energy density in the equivalent of the energy density in the curvature of space decays away slowly. Why would it be the case that we were born right at the right moment when these two things are comparable to each other? These two things evolve with respect to each other, the spatial curvature and the energy density. Wouldn't it be a coincidence if there we were at the right time to observe both of them?

This kind of argument convinced people that the spatial curvature must be zero; we must live in a flat universe. That turned out to be right. But, exactly the same set of words applies to the vacuum energy. The vacuum energy doesn't evolve at all as a function of its energy density. It doesn't change as the universe expands. It doesn't go up or down. But, the energy density in matter or radiation does go down as the universe expands. These two numbers change with respect to each other by quite a bit. However, we're inventing a universe now, claiming that if it's the data in which 30 percent of the energy density is matter, 70 percent is vacuum energy. Those numbers are close to each other. In the past, it was almost all matter. In the far past, it was almost all radiation. In the future, it will be almost all vacuum energy. Why are we lucky enough to be born at the right time when the vacuum energy is comparable to the matter in the universe?

That's the coincidence scandal. I actually don't have any good ideas for why that might be the case. So, we have a theory that fits the data. We keep getting more data; the theory keeps fitting. But, we recognize that the theory we have has holes in it, in the sense that we don't understand why certain parameters take on the values they do. That will encourage us to keep looking at more theories, at different possibilities. In the rest of these lectures, we're going to take some of these very seriously.

Lecture Sixteen
Smooth Tension and Acceleration

Scope:

We don't know much about dark energy, but we do know two things. First, it's smoothly distributed throughout space; otherwise, it would manifest itself in the dynamics of galaxies and clusters, which it apparently does not. Second, its density stays nearly constant as the universe expands; that's what makes the universe accelerate. Simple considerations of energy conservation imply that constant density is equivalent to negative pressure, or *tension*. Dark energy provides a persistent impulse to the expansion of the universe, which leads to cosmic acceleration in perfect accord with Einstein's general relativity.

Outline

I. Dark energy might seem mysterious and inscrutable—and it is. We don't know much about it, but we're not entirely ignorant.

   A. One thing we know is that the dark energy is persistent.

   1. It doesn't dilute away as the universe expands or, at least, not by very much.

   2. If it did (as matter and radiation do), it would have the same effect on the expansion rate of the universe as matter and radiation: It would cause deceleration.

   B. The other important feature of dark energy is that it is smoothly distributed.

   1. For whatever reason, the dark energy does not clump into galaxies and clusters.

   2. If it did, we would have detected it through lensing and dynamics.

II. Negative pressure is one of dark energy's characteristic features.

   A. The fact that dark energy is persistent actually implies that it has a negative pressure.

1. That is, if you had a box with nothing but dark energy inside, it would pull the sides of the box inward rather than pushing outward as an ordinary gas would.

2. This is less radical than it sounds; an example of negative pressure would be a series of rubber bands or springs tied to the sides of the box, pulling it inward. For this reason, negative pressure is sometimes simply called *tension.*

B. The relationship between persistence and tension is illustrated by the example of a piston with dark energy inside (and zero dark energy outside).

1. Imagine, in this purely hypothetical thought experiment, pulling out on the piston as a way of mimicking the expanding universe. The persistence of dark energy means that the energy density—the amount per cubic centimeter—stays the same as the volume expands.

2. Thus, the total amount of energy inside the piston increases; the volume goes up, while the energy per volume remains constant.

3. That means that you have to exert a pull on the piston to move it out—you are doing work in the process. In other words, the piston is pulling back on you, just as you would expect if the stuff inside the piston had a negative pressure.

C. One way of thinking about why dark energy makes the universe accelerate, rather than decelerate, is to point directly to the negative pressure.

1. According to Einstein, all forms of energy and momentum—including pressure—contribute to the deceleration of the universe.

2. If we worked through some math (which we don't do here), we would find that the source of the deceleration of the universe is the energy density plus three times the pressure. Why three? because there are three dimensions of space.

3. Therefore, if the pressure is sufficiently negative, the deceleration is negative (it's accelerating rather than

decelerating), causing things to be pushed apart rather than pulled together.

4. Some people like to refer to this as *antigravity*, but that's misleading; it's really just good, old ordinary gravity, with a funny kind of source.

D. Another way of thinking about why dark energy makes the universe accelerate—without any reference to negative pressure—is to say that the universe has a persistent density, which provides a continual source for spacetime curvature. This curvature, in turn, keeps the universe expanding.

III. Why does a persistent source of energy density make the universe accelerate?

A. Let's return to the claim that persistent energy density makes the universe accelerate.

1. In general relativity, the expansion of the universe is driven by the energy density.

   a. If the spatial curvature of the universe is zero, as data from the cosmic microwave background suggest, then the energy density is proportional to the expansion rate squared.

   b. That expansion rate is measured by the Hubble constant—by the Hubble parameter that relates the velocity that you observe in the galaxy to the distance that it is from you.

2. Therefore, a nearly constant density should lead to a nearly constant Hubble parameter. Why should a constant expansion rate be described as "accelerating"?

B. The point is that the expansion rate of the universe is not a velocity; it's the rate at which the universe increases in size by some multiplicative factor.

1. For example, imagine that the Hubble parameter takes some constant value such that the universe doubles in size every 10 billion years.

2. If we have two galaxies that are 10 million light-years apart, then 10 billion years later, they will be 20 million light-years apart.

3. But in another 10 billion years, they will be 40 million light-years apart, and 10 billion years after that, they will be 80 billion light-years apart!

4. Thus, a constant Hubble parameter means that the rate at which two galaxies move apart is actually getting faster and faster.

C. We can say the same thing in another way: Hubble's law says that the apparent velocity of a distant object is given by Hubble's constant times the distance—the Hubble parameter is telling you is how long it takes for the universe to increase in size by some fixed number. Now imagine that Hubble's parameter is truly constant.

1. With the current value of the Hubble constant, the universe would double in size roughly every 10 billion years.

2. For any given object, the distance is increasing as the universe expands; therefore, we would observe the recession velocity to be increasing, which is the definition of *acceleration*.

3. The only reason we expect deceleration in a conventional, matter-dominated universe is that the Hubble "constant" decreases sufficiently fast that the velocities decrease, as well.

Recommended Reading:

Goldsmith, *The Runaway Universe*, chapter 5.

Guth, *The Inflationary Universe*, appendix A.

Livio, *The Accelerating Universe*, chapter 5.

Questions to Consider:

1. Think of some everyday objects that have a negative pressure. What will their gravitational fields be like?

2. What would happen if the density of dark energy were increasing with time? How would we know?

3. Let your imagination roam free of the constraints of conventional technology and engineering (but not the laws of

physics). Is there any way you could imagine to use dark energy as a source of energy here on Earth?

Lecture Sixteen—Transcript
Smooth Tension and Acceleration

This is a lecture that I give only with some trepidation. This is the lecture in which we don't learn anything new about the universe, about dark matter, or dark energy. We just take something that we've already been saying about dark energy and try to really understand it at a deeper level. This is a lecture that you will not ordinarily hear about in speeches about dark energy in the accelerating universe because it's easier to say some words that sound like they make sense and then quickly move on to the next thing. So, the purpose of this lecture is to go deeply into those words that sound like they make sense, convince you that they shouldn't have made sense in the first place, but if you think about it hard enough, they begin to make sense again. So, I'm not sure whether this is even the right thing to do, but I think it's valuable enough that we really understand what is going on when we start talking about dark energy.

What we've said about dark energy is that it has two very crucial properties. One is that it is smoothly distributed through space. The same amount of dark energy is right here in this cubic centimeter as somewhere way in between galaxies and clusters very far away. At least the data are telling us that there is not substantially more dark energy inside the galaxies and clusters than in between the galaxies and clusters. That kind of makes sense. Had there been more dark energy inside a cluster, we would have noticed. We would have felt its gravitational field in the dynamics of the stuff inside the cluster.

The other part of dark energy is that it's persistent. The energy density in the dark energy is approximately constant as the universe expands. What that's telling us is that the dark energy is not made of some kind of particles that are becoming more and more dilute as the universe gets bigger. If that were the case, the energy would be going down. There'd be fewer particles per cubic centimeter. Whatever the dark energy is, it's something that stays the same as the universe gets bigger. So, that's something we'll have to get into. What are the possible candidates for stuff that could have a persistent energy density?

But, today, what I want to talk about is instead the idea that if you have a persistent energy density, it make the universe accelerate. Why it is that if the density of stuff doesn't go down, the manifestation of that in observable quantities is that the universe expands faster and faster. In fact, there are two different ways to explain this. They are just two different sets of words that we attach to the same set of equations. The truth is that this project of attaching words to equations is just because, as human beings, we like to have some intuitive grasp on what is going on. The equations themselves are completely unambiguous. There is no question about what does happen. The equations are telling you, absolutely once and for all, that if you have an energy density, which doesn't change as the universe expands, it will make the universe accelerate. It's perfectly clear at the level of the equations themselves.

But, we would like more than that. We would like to go beyond just being able to write some equations down to really have a deeper understanding of why it is like that. So, these are our attempts to attach words, to attach concepts that make sense to us, on to those equations. Sometimes, these attempts are going to necessarily be incomplete, or necessarily be fuzzy in some way. They will make a certain amount of sense to us, but they are not nearly as rigorous as the equations themselves. So, I'm going to give you two different explanations for why persistent dark energy makes the universe accelerate. Neither one is wrong, but you may find one explanation more compelling than the other one. That's perfectly OK. That's up to you.

The first explanation you will often hear is the following. I'll give you the whole thing and then we'll sort of unpack it and see why it makes sense. The explanation says: Dark energy has a negative pressure. In addition to positive energy density, there is also a pressure, but the pressure is a negative number. The thing that gravity responds to, according to Einstein, is a combination of energy density and pressure, and that combination added up can be a negative number if the pressure is sufficiently negative. Even if you have a positive energy density, you can still get a negative gravitational effect in the presence of negative pressure. That's what happens when you have dark energy and that's why dark energy makes the things in the universe move apart faster and faster. It's almost like anti-gravity pushing things apart.

You can decide for yourself whether or not that set of words I just said made sense. What I want to do is to go into them more deeply, try to understand why what we think of as dark energy would have a negative pressure, and why something with negative pressure would make the universe accelerate. Before we get there, we need to understand a little bit about the very notion of conservation of energy. Conservation of energy is one of the most cherished concepts in physics, going all the way back to Galileo, if not earlier. We want to know how it works in the context of relativity, which is after all a different theory than classical Newtonian mechanics. There is no necessary, logical reason that once we understand the laws of physics better, it needs to continue to be the case that our old cherished notions are still true.

In the context of relativity, the way that conservation of energy is manifested is different. In fact, you could say without being incorrect, that the energy in general relativity is just not conserved. I will try to make sense of that statement. You may have noticed that the energy of the universe seems not to be conserved in the presence of dark energy. On the one hand, I've said that dark energy has an energy per cubic centimeter that is approximately constant, maybe exactly constant. If the dark energy is vacuum energy, it is a strictly absolutely fixed amount of energy per every cubic centimeter.

On the other hand, the universe is getting bigger. The universe is expanding. There are more and more cubic centimeters in space as the universe expands; therefore, isn't it true that the total amount of energy in the universe is going up? Doesn't this mean that the energy is not conserved? The amount of energy is growing. So, the answer is yes; the energy is growing. It is not conserved. Then, you're supposed to say, isn't that bad? Doesn't that violate my cherished notions of conservation of energy? The answer is, it is not bad. It actually makes perfect sense within the context of general relativity.

So, to convince you of this, let me point out that even without dark energy, it is still the case that energy is not conserved in an expanding universe. Just think about an expanding universe that is more conventional. It has nothing in it but photons and ordinary matter particles, stuff that we certainly know exists. So, in a given region of the universe, this region is expanding. There are a number of particles in that region. The number of particles stays the same as the universe expands.

There are two different regimes for the kinds of particles we can consider. There are matter particles. Matter particles move slowly compared to the speed of light. There are radiation particles. Radiation particles move at or close to the speed of light. So, in each one of those cases, you get a different formula, a different idea of what the energy per particle is. For a matter particle, it's $E = mc^2$. For a slowly moving particle, most of the energy per particle is in its rest mass. What that means is that, as the universe expands, the energy in each individual matter particle stays constant. So, because this region of space is expanding, the total number of particles in that region is constant, the total energy in matter in that region will remain fixed, as the universe expands, just as you might hope it would do.

On the other hand, consider the energy in radiation in that same box. There, you have a fixed number of radiation particles. The number of particles isn't changing, but the energy per particle is going down. The effect of the stretching of spacetime is to increase the wavelength of every individual photon. The kinetic energy, if you like, of the radiation particles is diminishing because of the expansion of the universe; therefore, if you add up the total amount of energy contained in radiation in that box as it expands, you don't get a constant. You get a number that decreases as the universe expands. So, there is actually an amusing psychological effect going on here. With dark energy, as the universe expands, the total energy is not conserved because it's going up, and that bothers people. With radiation, as the universe expands, the total energy is not conserved because it's going down. That doesn't seem to bother people as much.

But, from the equation's point of view, both of those are exactly equally good or bad. If energy is conserved, the energy should not change, which means it should not go up or go down. So, the truth is that, in an expanding universe or in general relativity, more generally, there is no reason for the total energy of stuff in the universe to be conserved. By stuff, I mean not spacetime; I mean the substances that are in spacetime, whether they are dark energy, dark matter, ordinary matter, radiation, or what have you. If you were to dig down into the classical laws of physics, as Isaac Newton proposed them, there is a reason why energy is conserved.

The deep reason why energy is conserved was actually first understood by a mathematician named Emmy Noether. She figured out something called Noether's Theorem, which says that if there is a symmetry of nature, associated with every symmetry, is a conserved quantity. Something that is a symmetry of nature implies there is some number you can calculate that never changes. It doesn't quite go the other way. Just because something is a conserved quantity doesn't mean it comes from a symmetry, but it is very often the case. Energy, for example, is conserved because of a certain symmetry of nature. What symmetry of nature is it? It is time translation invariance. It's the statement that the laws of physics and the playground on which physics happens don't change as time goes on. That is a true statement in Newtonian mechanics. In Isaac Newton's universe, space and time are fixed and constant. Space does not change. Nothing happens to space and the laws of physics also remain unchanged. Therefore, in Newtonian mechanics, you can derive. You can go through a set of equations, which lead you to the conclusion that energy must be conserved. The general relativity works differently. General relativity allows space and time to be dynamical. In particular, we know that in cosmology the universe is expanding. In other words, the playground, the stage on which physics plays itself out is not invariant under time. The universe in the past was different than the universe in the future; therefore, the deep reason we had to believe in energy conservation is no longer true. It is not deep down a surprise that in an expanding universe, energy is no longer conserved.

Nevertheless, there is still an understanding of what happens. It's not that chaos is broken loose because the universe is expanding. There used to be, in Newtonian mechanics, a rule that says the total energy is constant. Now, in general relativity, there is a new rule, but there is still a rule. The new rule says how the energy changes, as the universe expands. If you tell me exactly how the universe is expanding, I can tell you exactly how the energy will change in response to that. Fortunately, there is actually a very easy way to understand this rule in terms of a much more mundane system, namely a piston. It's a piston in exactly the same sense in which you have pistons in your car, in the engine that is turning the energy in your gasoline to the kinetic energy of your car. In a piston, we imagine the simplest possible case in which we have a piston pushing into some substance and we're changing the volume of that substance by pushing the piston in or by pulling it out.

If you have ordinary gas—by gas, I mean the state of matter, not gasoline—inside your piston, if you have a gas like air or anything like that, it takes energy to push it in. You can get energy by allowing it to come out. We say that the gas in the piston has a positive amount of pressure. The gas in the piston is pushing on the piston; therefore, if we just let it go, we could hook it up to a little engine, which is what you actually do inside your car. You're getting energy out of the piston by allowing it to expand. That's what a positive pressure does. A positive pressure says you can extract energy by increasing the volume.

That's exactly what happens to a gas, for example, made of photons. In the universe, we can imagine applying the phrase gas to a collection of photons bouncing off the walls of our little piston. Photons are going to bounce into the piston. They are going to exert a force on it. That force is what we perceive as the pressure. So, the photons in the piston will push on the piston. We can extract energy by increasing the volume. That's just what the universe does. The universe, by expanding, takes energy away from those photons. If, instead, inside the piston we had a bunch of particles that weren't moving, particles that were motionless like matter particles, then we wouldn't get any energy by pulling on the piston. Again, that's exactly what happens in the universe. As the universe expands matter particles don't lose any energy in that way.

So, positive pressure means that, as we increase the volume, we take energy away. Therefore, you might guess that what negative pressure means is that we increase the volume, we put energy in. That's what the negative pressure is. So, if you try to imagine some physical system inside the piston that would have a negative pressure, it's something that when you pull on the piston, the system pulls back. For example, we could imagine a complicated system of rubber bands or springs inside the piston that were tied to the walls. If there is a rubber band that goes from one end of the piston to the part we're pulling out, when we pull, it will pull back on us. That is a negative pressure or, equivalently, it's called a *tension*. A rubber band has tension; it takes energy to make it bigger. It's not giving us energy when we make it bigger.

So, now, let's think about what if we had a piston full of dark energy? This is a special kind of non-physical thought experiment. We're imagining we have a piston with dark energy inside and no dark energy outside. Outside we just have zero energy everywhere. Inside, we have a system with the property that the amount of energy in every cubic centimeter is a constant. What happens? If we take that piston and we try to expand it, we're increasing the volume inside the piston. The energy per cubic centimeter inside remains constant, so the total amount of energy inside the piston goes up. In other words, we have to put energy into the system to pull out our piston. That is, if it makes sense, the proof that dark energy has a negative pressure. Dark energy is a system that requires energy for you to make it bigger. You need to put energy into the system somehow.

Since negative pressure is sometimes called tension, I've occasionally semi-seriously argued that dark energy is not a good name for dark energy. Everything in the universe has energy and there are lots of things that are dark, so the essence of dark energy is not really correctly described by calling it dark energy. The important things about dark energy are that it is smoothly distributed and that it has a negative pressure, or a tension. So, I proposed that we call dark energy *smooth tension*, which is both more accurate and kind of sexier than dark energy. It did not catch on, though. I was too late in coming up with this.

So, everything hangs together in this picture of how energy works. If you have an object or a physical system that has tension, that has negative pressure, then when you expand the volume that it takes up, you're putting energy into it. Contrary wise, if you have a system whose energy remains constant, you know that it has a negative pressure. However, in the case of the piston, there was an external agent. There was an outside without any dark energy, someone pulling on it. There is no equivalent in the case of the actual universe. There is nothing outside the universe pulling on it or pushing on it. The universe is just evolving, in accordance with Einstein's equations. So, the analogy breaks down a little bit there. If you counted for the piston the energy inside the piston and the energy of the person pulling on it or pushing on it, that total energy would be conserved. As far as we know, according to our current theories, anyway, there isn't anything outside pushing or pulling on the universe. It's just that the universe has an energy of its own that is not conserved.

We have to learn to deal with that. Let's see if that helps us understand why the universe is accelerating. Let's grant that dark energy is associated with negative pressure. Dark energy is something that it takes energy to make the volume bigger and bigger—so what? We have to go back to Einstein's equation, which tells us how the curvature of spacetime responds to stuff. Einstein's equation has a left-hand side involving the curvature of space and time. It has a right-hand side involving what we call the energy momentum tensor. In other words, Einstein's equation of general relativity involves a unification of different concepts, just like special relativity does. Remember that special relativity was inspired by Maxwell's theory of electromagnetism, which unified our description of electricity and our description of magnetism. Special relativity itself unified our idea of space with our idea of time into one notion of space time.

General relativity unifies the stuff that causes a gravitational field. It used to be, according to Isaac Newton, the stuff that caused gravity was mass. So, Einstein says in special relativity that mass is a form of energy. In general relativity, he says the stuff that causes gravity is every form of energy. It's not just mass; it's momentum. It's pressure; it's strain. It's a whole bunch of different ways in which energy can manifest itself. So, the thing that appears in the right-hand side of Einstein's equation is not just the energy density; it's also the pressure. If you work through some math, which we can't quite do right now, you find that there is a rough guideline that says that on some cubic centimeter of space, the thing that makes it expand or contract according to gravity is not just the energy density. It's the sum of energy density and pressure. The particular sum that it is, is the energy density plus three times the pressure. Why is it three? It turns out because there are three dimensions of space. If you lived in a universe that had five dimensions of space, the force of gravity on a cubic centimeter would be the energy density plus five times the pressure. What that three means is that if you have a pressure that is equal but opposite to the energy density, then it wins in that formula because there is a three multiplying the amount of the pressure. If you have an energy density that is positive, a pressure that is negative and comparable to the energy density, then the energy plus three times the pressure will be a negative number. That means that instead of pulling space together like you might guess, a sufficiently negative pressure pushes space apart. It makes the universe accelerate.

It's kind of like anti-gravity, in the sense that things move apart from each other under the influence of gravity rather than coming together. I could probably stop there and you would be willing to buy it. Let me just point out a tiny little slight of hand that happened in that argument. It's not a lie, it's not misleading you, but there is something that goes by very quickly, which is worth paying attention to. The gravitational effect of the pressure is what is being talked about here. A negative pressure, remember, inside the piston—what does a negative pressure do? It pulls on the piston. A negative pressure wants the piston to decrease the volume because that saves energy. But, what we're saying here is that a negative pressure in the universe makes the universe accelerate. How did that happen? What just went on?

What went on is if there is pressure in every direction, nothing happens. There is no force of the air pressure in this room on my hand because it's happening equally and oppositely on both sides of my hand. I don't feel any net force due to the pressure of the air in this room, even though the amount of pressure is 15 pounds per square inch. If I only had that pressure on one side, it would be pushing my hand over like that. The same thing happens with the negative pressure of the dark energy in the universe. It's exactly the same at every point in every direction; therefore, you feel precisely nothing from the direct impact of the negative pressure of the dark energy.

On the other hand, there is a gravitational effect. There is an indirect influence of the negative pressure due to its impact on the curvature of spacetime, and that effect is to make space expand faster and faster. So, all those words are true, they hang together, they make sense, but you need to buy into the claims that dark energy has negative pressure. There is a certain formula for the expansion of space, which is the energy density plus three times the pressure, and the fact that the pressure itself exerts no net direct effect.

Therefore, I can't resist giving you my favorite explanation for exactly the same phenomenon. Why does a constant energy density make the universe accelerate? I will explain why, without referring to the concept of a negative pressure even once. So, here is the other explanation, which is equally good, but I like it better. The explanation says that dark energy is persistent. The energy density does not go away. It remains constant as the universe expands; therefore, the energy density gives a constant persistent impulse to the expansion of the universe. That persistent impulse in every cubic centimeter of space manifests itself as *acceleration*. That's more or less the set of words I said already before. Let me unpack it a little bit more, so you understand the deep meaning behind that chain of logic.

We need to go back to the Friedmann equation, the equation in cosmology that relates the curvature of spacetime to the energy density. But, now, to save ourselves a little bit of conceptual work, we will set the spatial curvature equal to zero. We have data from the microwave background that says the spatial curvature is zero, so it's good enough to look at the Friedmann equation without the spatial curvature term. In that case, we get a very simple relationship: The energy density is proportional to the expansion rate squared. That expansion rate, of course, is measured by the Hubble constant, the Hubble parameter that relates the velocity that you observe in the galaxy to the distance that it is from you.

So, if the universe had nothing in it except for dark energy, just to do a simple thought experiment, what would happen? We would have an energy density that was constant. If you have a constant energy density, and the energy density is proportional to the Hubble parameter squared, then the Hubble parameter is constant. In conventional cosmology, when matter and radiation are important in the early universe, the Hubble parameter was much bigger at earlier times than it is today. In this fake toy universe we're discussing just for the moment, with no spatial curvature, no matter, and no radiation, nothing but dark energy, you would have a Hubble parameter that was truly constant, that never changed as the universe expanded. So, you're allowed to ask, wait a minute, I thought you just told me that the dark energy made the universe accelerate. How can it be that a universe where the Hubble parameter is constant is accelerating? In an accelerating universe, shouldn't the expansion rate be going up? The answer is no, and this is just one of those miracles of non-Euclidian geometry that manifests itself in general relativity. A constant expansion rate corresponds to an accelerating universe. That's because the expansion rate is not a velocity. The expansion rate, the Hubble constant, is not giving you exactly the velocity of a galaxy. It's telling you, if you know the distance to a galaxy, what would the velocity be?

We return to Hubble's law, Hubble's very simple equation, which says that the velocity of a distant galaxy is the Hubble constant times the distance. The Hubble constant relates all sorts of different galaxies to the velocities you observe. Now, ask what happens. According to Hubble's law, if the Hubble constant is strictly constant, it's not changing. Look at one galaxy and let the universe expand. The velocity you observe is the distance to that galaxy times the Hubble constant. But, as the universe expands, the distance is increasing; therefore, if the Hubble constant doesn't change, the velocity that we perceive will also increase. That's what we really mean when we say the universe is accelerating. We say that if you look at one galaxy and follow its velocity as a function of time, that velocity goes up. The galaxy moves away from you faster and faster. That's what it means to live in an *accelerating universe.*

In a *decelerating universe*, in a universe that only had matter and radiation in it, the distance would increase, but the Hubble parameter would decrease even faster. That's why you would see the velocity of a distant galaxy decrease in a decelerating universe. The statement that the universe is accelerating is equivalent to the statement that the Hubble parameter is either constant or decreasing more slowly than the distance is increasing. That's what it means to accelerate.

Let me say exactly the same thing in a different set of words. One way of thinking about what the Hubble parameter is telling you is how long it takes for the universe to increase in size by some fixed number. So, let's say the Hubble parameter is telling you it takes 10 billion years for the universe to double in size. Let's furthermore say that the Hubble parameter is constant. This is more or less what we have in our current universe. That's saying every 10 billion years, the universe doubles in size. What that means is if we take two galaxies that are 1 billion light-years apart, we wait 10 billion years, and they will now be 2 billion light-years apart. Another 10 billion years from now, they will be 4 billion light-years apart. In another 10 billion years from now, they will be 8 billion light-years apart. It's just like they told two friends and they told two friends. These galaxies move apart at a velocity, an apparent velocity, which is ever increasing and the reason why is that every cubic centimeter of space is expanding at a constant rate. A constant expansion rate of space plus an increasing amount of space in between the galaxies leads to an acceleration of the two galaxies away from each other.

That is my favorite explanation for why dark energy makes the universe accelerate. You don't need to go through the intermediary of a negative pressure, which is kind of a difficult-to-grasp idea all by itself. You can just say that what energy does is contribute to the curvature of spacetime. In a flat universe, what that means is that energy contributes to the expansion rate of spacetime. If that contribution doesn't go away, the expansion rate of spacetime will persist. If we lived in a universe without dark energy, a universe with only ordinary matter and radiation, the far future of the universe is one in which it expands ever more slowly. It gets emptier and emptier, and approaches exactly the situation you would have if there were no stuff in the universe—if you were not expanding whatsoever, if you were in a static, un-expanding, empty spacetime. But, with dark energy, with stuff that doesn't go away, there is always something that is making the universe expand with this constant magnitude. The manifestation of that to us is that we see individual galaxies moving away faster and faster.

Hopefully, all of that now makes perfect intuitive sense to you, except we're still left with the question, what is this dark energy stuff? With the next lecture, I'll be wearing a different tie, and we will start thinking seriously about different candidates for what the dark energy might be.

Lecture Seventeen
Vacuum Energy

Scope:

Dark energy is smooth and persistent, but what exactly *is* it? There are several possibilities, the simplest of which is *vacuum energy*—an absolutely constant amount of energy inherent in every cubic centimeter of space itself. Vacuum energy is equivalent to Einstein's idea of the *cosmological constant*, first proposed by him and later repudiated as his biggest mistake. Vacuum energy provides a good fit to the cosmological data but at the cost of a deep mystery: Why is it so small? Simple estimates imply that, if there is any vacuum energy at all, there should be enormously more than we observe. This *cosmological constant problem* remains one of the deepest mysteries of modern physics.

Outline

I. The simplest possible kind of dark energy is vacuum energy.

    A. We know that the dark energy is fairly smoothly distributed and fairly persistent in density. The simplest possibility, therefore, is that it is exactly constant in density, both across space and through time.

        1. This possibility is called *vacuum energy*, or the *cosmological constant*.

        2. The concept of vacuum energy is not exactly the same as that of dark energy; the former is a specific candidate for what the latter might be. (In other words, the vacuum energy might be exactly zero, and the observed dark energy might be something else, as we'll discuss in the next lecture.)

    B. Vacuum energy is simply the energy inherent in space itself.

        1. Consider a tiny region of space that is perfectly empty: no matter, no radiation, nothing at all.

        2. Then ask, "How much energy is contained in this region?" You might think that the answer is necessarily zero, but that's not right; according to general relativity, the answer is a constant of nature—the energy of empty space.

II. Einstein thought of the cosmological constant as his greatest blunder.

   A. Soon after Einstein invented general relativity, he applied it to the entire universe.

   1. He found that such a universe must be either expanding or contracting—there was no static solution, because the matter in the universe would be pulling on all the other matter.

   2. But astronomers of the time (before Hubble came along) assured him that the real universe was, in fact, static.

   B. Einstein could have stuck to his guns and made the bold prediction that the universe must be either expanding or contracting. But instead, he invented the concept of vacuum energy, which he called the cosmological constant.

   1. He invented a model of the universe in which the accelerating impulse of the cosmological constant precisely balanced the decelerating impulse of the matter (with some spatial curvature added to make the math work out correctly).

   2. This was not Einstein's best idea, because the resulting model wasn't really stable; any tiny perturbation away from perfect balance would make the universe either expand or collapse.

   C. When Hubble reported that the universe was actually expanding, Einstein realized that he had made a mistake and reportedly referred to the cosmological constant as his "greatest blunder." However, Pandora's box is not so easily closed; once the concept of an energy density inherent in empty space is introduced, it becomes hard to ignore.

III. Vacuum fluctuations contribute to the value of the cosmological constant.

   A. The modern picture of the vacuum relies heavily on the precepts of quantum mechanics.

1. In particular, we have Heisenberg's uncertainty principle, which says that it is impossible to pin any system into one particular well-defined state.

2. For example, we can't guarantee that a pendulum is sitting at the bottom of its arc, even if we know that it's in the lowest-energy state. There is always some irreducible quantum jitter.

B. Usually, we apply the uncertainty principle to an elementary particle, such as the electron, but it also applies to empty space. In that context, it means that it's impossible to make space perfectly empty—even in the emptiest state, there will be *virtual particles* that constantly pop in and out of existence.

C. The existence of virtual particles is not in doubt—not only are they predicted by quantum mechanics, but their effects have been observed, for example, in the energy levels of atoms. But remember, everything couples to gravity, and that includes virtual particles. The energy of these particles contributes to the vacuum energy (as do other things).

D. We can't precisely calculate the predicted vacuum energy from quantum fluctuations, but we can roughly estimate its likely value.

1. Our best estimates, unfortunately, give an answer (10^{112} ergs per cubic centimeter) that is much larger than the observed density of dark energy (10^{-8} ergs per cubic centimeter) by a factor of 10^{120}. That's 1 followed by 120 zeros, an absurdly large number. (For comparison, there are only about 10^{90} particles in the entire observable universe.)

2. This is known as the *cosmological constant problem*, and nobody knows how to solve it.

E. The cosmological constant problem was known long before dark energy was discovered. In those days, however, there was a plausible loophole: Some unknown symmetry or other law of physics might be setting the vacuum energy to zero for reasons that we would someday discover.

F. If the observed dark energy is, in fact, vacuum energy, those hopes have been dashed.

1. This makes the problem worse rather than better; it's one thing to imagine an unknown mechanism that completely cancels some number and sets it to zero, but it's much harder to imagine a mechanism that makes the number much smaller than it should be but not quite zero.

2. So in the next lecture we'll consider an alternative scenario: Perhaps the vacuum energy is zero after all and the dark energy we're seeing is, not a cosmological constant of the vacuum, but instead something dynamical.

Recommended Reading:

Cole, *The Hole in the Universe*, chapter 4.

Livio, *The Accelerating Universe*, chapter 5.

Vilenkin, *Many Worlds in One*, chapter 2.

Questions to Consider:

1. Put yourself in Einstein's shoes after he had invented general relativity but before Hubble had discovered the expansion of the universe. Should Einstein have been able to predict that the universe would be expanding or contracting? Was the cosmological constant really a great blunder?

2. Can empty space really have energy? Would there be any way to create a region of space with even less energy than the vacuum energy?

3. Given that the cosmological constant is much smaller than it should be, do you think it was a sensible expectation for physicists to have guessed that it was exactly zero, even if they didn't know why?

Lecture Seventeen—Transcript
Vacuum Energy

We have every reason to be proud of what we've learned about the universe so far, both over the course of these lectures and over the course of the past hundred years as working physicists and cosmologists. We know that the total energy density of the universe is about the critical density, what you need to make the universe spatially flat. We know that of that density, it's about 5 percent ordinary matter, 25 percent dark matter, and 70 percent dark energy. We have a great theory for what the ordinary matter does. We have the Standard Model of Particle Physics, which tells us what the particles are and how they interact, consistent with all the experiments we've done here on earth so far.

We don't know what the dark matter is, but we have more than enough different candidates. It's not a surprise to us or a deep mystery how you could get dark matter to be 25 percent of the universe. We have lots of different ideas for what it could be. Furthermore, we have very good ideas about how to test those theories, ways that we can create the dark matter and particles individually, and also to detect it coming from outer space here to our detectors on earth. Now is the lecture on which we face up to the fact that the dark energy is not so simple. The dark energy is something about which we do not have very good ideas about what it might be. We're going to spend three lectures looking at different possibilities for why the universe is accelerating and spatially flat, why we seem to believe that there is dark energy, at the end of which you will realize that none of the theories are especially promising. None of them are like the case of supersymmetry in the case of dark matter, where you have a theory that does other things, naturally provides you with a candidate, and whether it's right or not, is at least a very plausible scenario.

Let's think about exactly why we think that there is dark energy. We have a set of evidence for it. That evidence is very different than the evidence that we have for dark matter. The dark matter evidence comes from the local dynamics and behavior of stuff in the universe attracting other stuff. We have galaxies, clusters of galaxies, gravitational lensing, and the growth of structure in the universe. All these say there is more stuff here in those bound systems than there is outside. If we naturally attribute that to cold, non-interacting massive particles, we seem to fit the data.

Dark energy, on the other hand, is found globally. We look at the acceleration of the universe due to the kind of "stuff" that is inside it, we look at the spatial geometry of the universe, which is a way of measuring everything all at once. We find that if we imagine that 70 percent of the energy density of the universe is smoothly spread throughout space, taking the same amounts inside clusters and galaxies and outside, and also persistent as a function of time, not redshifting or diluting away as the universe expands, we can fit all these data all at once. So, we call this mysterious new substance dark energy.

What could it be? The two things we know about dark energy are that it's smoothly distributed through space and nearly constant in time; therefore, certainly the simplest thing it could possibly be is something that is absolutely 100 percent the same from place to place in space and the same from moment to moment in time. The data are not yet telling us that that is precisely the case. The data are, of course, consistent with some slop in that conclusion. There could be small variations from place to place. So, that's going to be the subject of the next lecture. This lecture is going to be about the possibility that it really is truly constant. The physical underpinnings for those two possibilities are very, very different, even though their observational signatures are pretty much similar.

We should say, before we go there, that even though we're going to contemplate in this lecture the possibility that empty space has an absolutely constant energy known as the *vacuum energy*, the vacuum energy is a candidate for what the dark energy could be. These are not synonyms with each other. The dark energy is a label given to whatever it is that is out there in the universe that is smoothly distributed and persistent, 70 percent. The vacuum energy is a specific idea for what it might be, the energy of empty space itself. It's still possible that what we call the vacuum energy is zero and what we call the dark energy that is making the universe accelerate is something else entirely. But, for this lecture, we're going to concentrate on the possibility of vacuum energy.

What do we mean by vacuum energy? We mean the energy of empty space itself, but what does that mean? That means that you take a little region of space, a little cubic centimeter, and you remove from it everything you can possibly remove from it, so it's completely empty. You remove all the ordinary matter, all the dark matter, all the radiation, all the neutrinos—there is really literally nothing there—and you ask, how much energy is there contained inside this cubic centimeter? In the context of either general relativity, our best theory of gravity, or quantum field theory, our best theory of microscopic physics, there is no reason why the answer to that question should be zero. There is some number, some constant of nature, which tells us how much energy there is in every region of empty spacetime.

This picture that you're seeing here is actually an artist's impression, my impression, of what dark energy, vacuum energy, is like. You can think of it as a false-colored image of empty space itself. Of course, empty space is really completely dark and invisible, completely see-through and transparent, so this is just an attempt to make you contemplate the idea that even in this little bit of empty space, there is still energy there. The energy is not contained in some substance that is located there and could be moved around. It's inherent in the fabric of space and time itself. That is the idea of vacuum energy—the bare minimum amount of energy you can fit anywhere. There is nothing you could do to that cubic centimeter, no physical process, that would lower the energy below that.

What do we mean by the energy of empty space? We're saying over and over again the same words. It's the energy that you have in an empty cubic centimeter of space, but you're still sitting there thinking, what does that really mean? To a physicist, when you say, what does that mean, you're asking, what does it do? How would we know that it is there? Is there some operational, observable consequence of saying these words? Now we get a little subtle point, actually, because in almost all areas of physics, the actual value of the energy doesn't matter. What matters is how much energy changes when one process happens, the energy that goes up or goes down or the energy that gets exchanged between two different ways in which it can manifest itself.

For example, you might be familiar, from high school or college physics, with experiments like an inclined plane, a ball or a box rolling down some inclined plane, and people will talk about the energy that is contained in this ball rolling down the plane. It will have potential energy, depending on how high it is, and kinetic energy, depending on how it's moving. If everything is very smooth, there is zero change in the total energy. You're just converting potential energy into kinetic energy. Here's the thing—if you ask what the total amount of energy of that ball rolling down the plane is, it doesn't matter what the answer is. If you did exactly the same experiment from the plane being right here, the ball rolling down, or raising the plane up so it's up here, the potential energy of the ball would change, but the way in which the energy goes from being potential energy to kinetic energy wouldn't change. Of course, the motion of the ball down the plane would be exactly the same, whether it was up here or whether it was down here. As long as the tilt of the plane is the same, the local physics is exactly identical. The total amount of energy that the ball has is completely irrelevant.

You might say, even if there was empty space energy, vacuum energy at every point in spacetime, how would we know? The answer is that gravity is the one thing in nature that really does care about the absolute amount of energy. Remember, Einstein tells us that gravity responds to everything. Gravity is universal. And so, what's going to turn out to be the case is that the amount of energy that is contained in empty space doesn't affect anything about, for example, the Standard Model of Particle Physics. The only thing it affects is gravity. It acts to curve spacetime; it creates a gravitational field. That's how you can detect it. That's how, in fact, we're claiming that we did detect something like it, by looking at the curvature of space and, of course, the acceleration of the universe.

What is the effect of this vacuum energy? We've already said that it makes the universe accelerate. If there is some energy density in empty space, that energy density doesn't go away as the universe expands. It's a constant. It's a constant of nature from place to place and from time to time. So, as the universe is expanding, this vacuum energy is giving a perpetual impulse to the expansion of space, and we perceive that as the acceleration of the universe.

There are other effects. If you had infinitely good measuring apparatuses, you would be able to detect the existence of the vacuum energy using all sorts of gravitational experiments. The classic experiment that lets us test general relativity in the first place is the motion of the perihelion of Mercury. The planet Mercury moves around the sun in an ellipse and, according to Isaac Newton, that ellipse should be fixed in orientation once and for all. If you put aside perturbations from the other planets and imagine a perfect solar system that just has the sun and Mercury, that ellipse is exactly the same for all time. Einstein comes along in general relativity and says that that ellipse should change slightly. That was actually already known to be the case. Before Einstein came up with this theory, astronomers had already determined that Mercury was slowly shifting in its orbit. That is a way in which we were able to tell that general relativity was a better theory than Newtonian gravity.

It turns out that if you had a vacuum energy, you would also add a small new effect to the motion of the planet Mercury. However, that effect, when I say small, is really, really small. There is no possible way we will ever be able to test the possibility of vacuum energy by looking at Mercury. It's just swamped by things like me sneezing here on earth. There are a million bigger effects than the vacuum energy when you're thinking about how Mercury moves around the sun.

The reason why we can detect the vacuum energy in cosmology is that because the effects accumulate. When you're looking at the motion of a galaxy, or the supernovae in some galaxy, there are many, many cubic centimeters between us and that galaxy. Every one of those cubic centimeters is expanding just a little bit faster because of the presence of vacuum energy. It's that accumulated effect over cosmological distances that enables us to actually detect the fact that the vacuum energy is there.

The first way that physicists started thinking about this concept of vacuum energy was actually going back to Einstein himself. Einstein, soon after inventing general relativity, started thinking about cosmology. He knew that he had a very different way of thinking about space and time than Newton did. Isaac Newton had a picture in which space and time were both fixed and absolute. Einstein has a picture in which space and time have a geometry. They are dynamical and they can change. So, he started thinking about the entire universe. As a simplifying assumption—we're now talking 1916, 1917, 1918, long before we knew the universe was expanding—Einstein told us the simplifying assumption that the universe was pretty much the same everywhere. He found that, within that possibility that the universe was pretty much the same everywhere, he was not able with his equation governing the curvature of space and time to find a solution in which the universe was static, in which it was smooth everywhere and more or less staying the same as time went on.

We know today the reason why—because the universe is, in fact, expanding. It is not static. That makes perfect sense. If you have galaxies spread throughout the universe, they are going to pull on each other. In the context of Newtonian gravity, if you try to ask the question, what could happen if you filled space full of galaxies the same number everywhere, would they pull on each other or not—it's very hard to get an answer. On the one hand, you might think that yes, they are pulling on each other; therefore, they should come together. But, there is an equally plausible looking analysis that says, here I am, this galaxy. Every other galaxy is pulling on me, but they are equally distributed. The net force is zero and I don't move. Nothing moves. Within general relativity, there is a different answer that says, you can think about nothing moving, but space itself can get bigger or smaller. So, in general relativity, if you have nothing but matter in the universe, the answer is unambiguous. Space better be either expanding or contracting.

There is a little bit of a false story that goes around about Einstein thinking about exactly these issues. It says, Einstein was blinded by his philosophical presuppositions. He wanted to believe, this false story goes, that the universe is static for some religious or philosophical reasons. Therefore, when he found that his equations didn't describe a static universe, he was upset and tried to change them. The truth is that he did believe the universe was static, but it wasn't because of a philosophical predisposition. It's because he asked his astronomer friends. At that time, in the 19-teens, in between 1915 and 1929, as far as anyone knew, the universe was static. That was what the observations were telling us. It wasn't until Hubble, in 1929, discovered the fact that there is a relationship between the velocity of distant galaxies and their distance, did we realize that space is, in fact, expanding and dynamical.

So, what Einstein was trying to do was to fit the data. He thought the universe was static. He showed that his own equations did not allow for a static solution, if you just imagined the universe full of galaxies, or even full of matter of any sort. So, what did he do? In 1917, he changed his equation for general relativity. He added a new term, which he called the *cosmological constant*. These days, we consider changing Einstein's equations to be an interesting step, but a dramatic one. We have very good success with those equations. But, you have to remember that Einstein himself was coming up with new equations all the time. In fact, the final version of his equation was not anywhere near the first version of it. He was changing it to fit things both philosophical and experimental all along.

Finally, in 1917, he said, "Well, I can add a term to the left-hand side of my equations." So, here are, of course, Einstein's equations. On the left-hand side, you see the term that tells you how curved space and time are. On the right-hand side, you have the term representing stuff in the universe, energy and momentum and so forth. Einstein put a new term on the left-hand side for which he could then find a solution that the universe could be filled with galaxies and yet not expanding or contracting. The solution that he called the cosmological constant we now know that you can move to the right-hand side and it's exactly the same as vacuum energy. In fact, there are some people who like to argue whether or not this term deserves to be on the left-hand side with the spatial and spacetime curvature bits or deserves to be on the right-hand side with the energy bits. Of course, the truth is that it absolutely makes no difference whether a term in an equation is on the left side or the right side. It functions in exactly the same way and has the same interpretation. Einstein would have called it the cosmological constant. We will call it the vacuum energy, but it is the same thing.

So, vacuum energy acts to make the universe accelerate. It basically pushes things apart as the universe expands. Ordinary matter pulls things together. Basically, all Einstein did was find a solution where those two effects exactly balance. Now, it's true he could find a solution where the universe was not expanding or contracting, but of course this solution was unstable. If you perturbed it a little bit, if things just started moving together a tiny bit, the pulling together force would win and it would collapse. If things started expanding a little bit, the pushing apart force would win and things would start accelerating. Einstein found a solution that fit the data, but it wasn't a physically plausible solution; it wasn't something that was robust to small changes in what happened.

Then, of course, in 1929, Hubble comes along and says, "Well, the universe is not static. It's expanding, just as Einstein could have predicted in 1917." According to George Gamow, who later was one of the primary thinkers about the Big Bang model, Einstein later claimed that his "biggest blunder" as a scientist was to introduce the cosmological constant into his equations. If he had not done that, he would have made a prediction. He would have been able to say, "Despite what the current observations say, circa 1917, I predict that the universe should either be expanding or contracting." But, he blinked and he wasn't able to do that. He then said, "Away with the cosmological constant. I'm sorry I ever invented it."

Once you do that, you can't undo it. Pandora's box is not so easily closed. These days, we realize that what Einstein called the cosmological constant is what we call the vacuum energy. We can start asking questions like, if there can be energy density in empty space, how much should there be? Can we make some sort of reasonable estimate for how much energy there should be in every cubic centimeter of space, the minimum possible value? It's a very fuzzy story that goes back and forth. There is no precise answer to this question, what should be the vacuum energy according to our current theories? The true answer is it can be anything at all. It is a constant of nature. It's like saying, what is the mass of the electron? There is no ahead-of-time answer to what it should be. You have to go out there and measure it.

On the other hand, there are reasonable answers and unreasonable answers. We can do a pretty good job of estimating order of magnitude, how big the vacuum energy should be. We have both what we would call a classical contribution to the vacuum energy, the thing that it should be if you ignored all of quantum mechanics, and then, as we'll explain, quantum mechanics adds new contributions on top of that. Together we get an estimate. We can compare it with what we see. The answer is, it's nowhere close. The estimated value of vacuum energy is much, much larger than what we observe. Let's think deeply about why that is true.

First, we need to talk a little bit about quantum mechanics. Quantum mechanics is the correct theory of the world, as far as we know—at least the correct framework in which to be thinking about the world. It replaces Isaac Newton's classical mechanics. In both classical mechanics and quantum mechanics, you have physical systems, and those physical systems do things. They evolve; they obey equations of motion. The real difference between classical mechanics and quantum mechanics is that, in classical mechanics, you can observe everything there is to know about the system. So, if I have a system that is a ball rolling down a plane or a pendulum swinging back and forth or a set of springs, I can measure where all the components of that system are. Then, I can use the laws of physics to predict where they will be in the future and where they were in the past, in principle, to arbitrary accuracy.

In quantum mechanics, meanwhile, there is a rule that says, you cannot measure all of the properties that a system has. In fact, you are dramatically unable to measure. Think about a coin, which you can flip. It will be heads or tails. While it's still spinning in the air, if you asked whether that coin is heads or tails, classically, it's somewhere in between. It's rotating in between being heads or tails. Quantum mechanics is like a coin that you flip and it's rotating in the air, but nevertheless, every time you look at it, it's either exactly heads or exactly tails. It's not that we don't know which one it is. It's that when you're not looking at it, it is some superposition. It is neither heads nor tails. It is described by some angle. But, when you look at it, you never see that angle. You always see it to be exactly heads or exactly tails.

For a more realistic example, think about a particle like an electron. An electron, classically you would say, has a position. It's located there and it has a velocity; it's moving in some direction. According to quantum mechanics, it's not that we don't know what the position is. It's that there is no such thing as the position of the electron. What there is, is a function of space called the wave function. The wave function tells us, if we look at the electron, where are we most likely to see it? What is the probability of finding it in different places? But, the true answer to where the electron is, is not a question that has an answer. There is no such thing. When you look at the electron, you always see it in a position, but when you're not looking, it doesn't have a position. It has a probability of having different positions. That is the origin of uncertainty in quantum mechanics. What you can observe is not what there really is, but what there really is, this wave function, tells you the probability of getting different answers to your observations.

Let's apply this to a very specific system, which is not exactly the real world, but it's a good analogy for the real world, and that's a simple pendulum—just a weight that can rock back and forth. It can be stationary, just sitting there, or it can be going back and forth with some amplitude and some frequency. This is a good classical thing. This is exactly the kind of thing that Galileo used to look at when he was falling asleep in the cathedral in Pisa, and he would time the pendulum with his pulse because he didn't have a very good clock back then. But, these days, what we've learned to do is to take classical systems and quantize them, to put them in a quantum mechanical framework and see what happens. In the classical pendulum going back and forth, it has energy. It has a potential energy, depending on how high it's gone up, and it has a kinetic energy, depending on how fast it's moving. Furthermore, those two things, the kinetic energy and the potential energy, can take on absolutely any value.

In quantum mechanics, on the other hand, two things happen. One is that the energy cannot take on any value. It takes on certain discreet values. That's why it's called quantum mechanics. There are specific energy levels that you can see. You can't see anything in between them. The other thing is if you ask what the lowest energy level is, you get a different answer in quantum mechanics than you would have in classical mechanics. In other words, all you're saying is, here's a pendulum. Put it in the zero-energy state. Put it in the state of minimum energy. Classically, it's kind of obvious what to do. You stop the pendulum from moving. You make it sit there, so it's located at the bottom of its trajectory. Its potential energy is minimized and there is no kinetic energy. Quantum mechanics says that that would be the same as precisely specifying the location of the pendulum, and you can't do that. Therefore, the minimum energy configuration still has some uncertainty in where it is and, therefore, still has some uncertain energy.

So, this idea of a pendulum is actually quite analogous to the quantum fields that we have in the real world. A pendulum has, classically, energy. Quantum mechanically, it has a slightly higher minimum energy than classically. What we have in the real world are fields oscillating through space and time. But, it turns out to be quite a good approximation to think of the field at every point in space as a little pendulum going back and forth. It has kinetic energy; it has potential energy. Quantum mechanically, it has a little bit more. But, I want you to notice two things about this analogy. First, if you ask what the energy of the pendulum is, just like the inclined plane, there is no right answer. The potential energy of that pendulum would change if we moved the whole apparatus up by a foot. But, none of the physics would change; none of the laws of motion of the pendulum would be altered in any way. The quantum fields in spacetime have exactly the same property. There is a minimum amount of energy that you can ascribe to them that is true classically, even in empty space, and it's a completely arbitrary number.

But, then, what quantum mechanics says is that on top of that completely arbitrary number, there is also some quantum mechanical jiggle. Quantum fields have the same property. In addition to the completely arbitrary answer to the question of what the minimum energy is, quantum mechanics adds an extra contribution to that minimum energy because of the fact that the fields are jiggling back and forth. This is just Heisenberg's uncertainty principle, the fact that you can't localize the quantum mechanical system with absolute precision. Because of that, there will always be quantum jitters. In the case of a field, like the electron field or the electromagnetic field or the gluon field, these jitters carry energy and they contribute to the energy density of empty space. They contribute to the vacuum energy, in other words.

There is a rule in quantum field theory. When you look at a field, what do you see? Just like for a single electron you have a wave function, when you look at it, you see a particle. For a quantum field, you see a collection of particles. A quantum field is basically a bookkeeping device that tells you how many particles there are all over the place. So, you might think that if I have a quantum field, its vacuum state, which to physicists means its lowest energy state, the state I can put the quantum field in that has the least energy I can possibly have, would have zero particles. If there are particles, then you have some energy there. The vacuum state of the field, the lowest energy state, should have no particles anywhere. It should just describe empty space. That is correct, in the sense that there are no particles in empty space that you can see.

Nevertheless, Heisenberg and his uncertainty principle are telling you there are particles that you can't see—we call these *virtual particles*. The fact that you can't pin down the quantum field to some precise configuration is telling us that you can't avoid virtual particles popping in and out of existence. These virtual particles are not a new kind of particle like we already have electrons and quarks, now we also have virtual particles. They are the good old particles that we know and love—photons, electrons and positrons, and so forth—fluctuating in the vacuum, fluctuating in empty space itself.

These virtual particles certainly exist. This is not some crazy hypothetical thing. It was a crazy hypothetical thing 70 or 80 years ago, but since then, we have detected the effects of these virtual particles. They interact with ordinary particles passing through empty space, and the effects of these interactions change the atomic energy levels in ordinary atoms, ordinary stuff. They give rise to corrections to formulas in particle physics, and these corrections have been tested to exquisite accuracy. In other words, we have a picture in quantum field theory of empty space in which empty space is not boring. Empty space is alive, popping with virtual particles and anti-particles that appear and then disappear. We can't see them directly; we can't pull a virtual particle out of the vacuum and make it real, but we can tell that they are there because they have effects indirectly on the behavior of other particles.

For example, they have effects on gravitons. These virtual particles have a gravitational field. They carry energy and that energy that they have is precisely the vacuum energy—a contribution, if you like, to the amount of energy density in empty space. What we said then is that just like for a pendulum, if you ask what the minimum energy of empty space is, there are two contributions. First, there is a completely arbitrary classical number. There is some number there, in the laws of physics, that says, if quantum mechanics weren't true, the vacuum energy would be such and such a number. We have no idea what that number is.

We also have a quantum mechanical uncertainty, a zero-point energy of the vacuum fluctuations of every field in the universe added on top of that. What is this energy? How much is the shift in energy from quantum mechanics? There, we can at least estimate it. So, even if we didn't know the classical energy, you might imagine that the quantum mechanical shift was of the same basic size as the original vacuum energy. It turns out that if you do a naive estimate of how much energy quantum fluctuations add to the vacuum, you get infinity. That's not right. But, in quantum mechanics, infinities happen all the time and we know how to fix them. We put in some cutoff, we stop including contributions from very, very, very short length scales where space time itself might dissolve into some sort of quantum foam.

Then, once we have that cutoff, we get a finite answer for what the quantum mechanical contribution to the vacuum energy is. In terms of numbers, it turns out to be 10^{112} ergs per cubic centimeter. That is a lot of ergs. An *erg* is a measurement of energy. If you ask, given the data, given the observations from cosmology, how many ergs per cubic centimeter are there, the answer is 10^{-8} ergs per cubic centimeter. In other words, our best guess, our best estimate on the basis of modern physics of how much energy there should be in the vacuum, is 10^{120} times bigger than the amount that we actually see. One trillion, trillion, trillion, trillion, trillion, trillion, trillion, trillion, trillion, trillion times bigger than what we actually see. That's a theory that does not match the data.

This is called the *cosmological constant problem* and is probably the largest unsolved problem in theoretical physics. We had known for a long time that there was a cosmological constant problem, long before we actually detected the vacuum energy. It was obvious that the vacuum energy was not nearly as big as our naive estimate said it should be, so we had that problem. But, before we detected the acceleration of the universe, the best idea was that there was some secret symmetry of nature, some secret mechanism that took this huge vacuum energy you should have and exactly cancelled it, made it equal to zero. Even though we didn't know what that symmetry was, it didn't seem like such a stretch that some day we would detect it.

But, now that we've found some vacuum energy, now that we've found some dark energy that we can attribute to the energy density of empty space, the problem has become much harder. We don't want to multiply the real number by zero; we want to multiply it by 10^{-120}. It's as if you see someone on the street who is flipping a coin. They flip a coin and they say, do you think it's going to be heads or tails, and you say heads. They flip it and it's tails. They flip it again and it's tails. They flip it 299 times in a row, and it's tails every time. You don't know why; you don't know what the mechanism is. But, if they then say, I'm going to flip it again, what do you think it will be, you're going to say tails. They flip it and it's heads. Something made something go away. Something made the vacuum energy disappear 2^{299} times, but then there is a leftover there, that 300th time.

This tiny amount of vacuum energy, by all rights, shouldn't be there or should be bigger. We're stuck with a situation where it's there.

So, it might be right. It might be that once we finally understand everything, we get a better formula for predicting what the vacuum energy would be, and we get the right answer. Right now, we're simply at a loss. If the cosmological dark energy is the energy of empty space, it's an idea that fits the data, but about which we have no understanding. Therefore, we're going to try other things. We still don't know why the vacuum energy is small, but maybe it is zero. Maybe the actual dark energy we're seeing is something dynamical. That's what we'll be exploring in the next lecture.

Lecture Eighteen
Quintessence

Scope:

Perhaps the vacuum energy really is zero, and dark energy is something else—some form of energy that isn't really constant. It could be slowly varying but nevertheless dynamical: an idea known as *quintessence*. Dynamical dark energy could result from a new field in nature, analogous to the electromagnetic field but remaining persistent as the universe expands. This is an intriguing idea, and physicists have been trying to fit it in with the rest of what we know about particle physics. Fortunately, it's a testable idea: We can use observations to constrain the dark energy *equation-of-state parameter*, which tells us how close to strictly constant the dark energy density really is.

Outline

I. Dark energy could be dynamical, rather than strictly constant.

    A. One motivation for considering dynamical dark energy is the cosmological constant problem: It is still easier to imagine setting the vacuum energy precisely to zero, even if we don't know how to do it, and then adding a new field to bring the observed dark energy up to its current value.

    B. We know from observations that the density of dark energy is nearly constant through space and time. Vacuum energy is the idea that it is precisely constant; however, we can also consider the possibility that there are mild variations in the density.

    C. The simplest way to get such variations is to imagine that the dark energy is generated by some dynamical field, called *quintessence*.

        1. This would be a new bosonic field in nature, somewhat similar to the electromagnetic field, but with the crucial difference that its energy density remains nearly constant as the universe expands.

2. It's not difficult to invent explicit particle physics models of such fields if one is allowed arbitrary freedom to tune the relevant parameters in the theory to the appropriate values.

3. If a quantum field is like an infinite collection of swinging pendulums, the quintessence pendulums are moving in cold molasses, taking billions of years to reach the bottom of their arcs.

D. Another motivation is the *coincidence scandal*: Why are the densities of matter and dark energy so similar? The important fact here is that the two quantities evolve rapidly with respect to each other as the universe expands: The matter density dilutes away, while vacuum energy remains constant.

1. If they are similar today, they were very different in the past (matter was dominant) and will be very different in the future (vacuum energy will be dominant).

2. Replacing a constant vacuum energy by a dynamical field allows us to contemplate some dynamical solution to this problem.

3. Once again, no promising solution along these lines is known, but at least we can contemplate it.

II. Quintessence comes in a variety of forms (all completely hypothetical, of course).

A. If the density is gradually decreasing, the universe can be accelerating, but the Hubble parameter will nevertheless gradually diminish. We can also, more dramatically, imagine that the dark energy density is actually increasing; this possibility is called *phantom energy*.

1. In that case, the Hubble parameter will increase. And if it increases quickly enough, we can have a *Big Rip*—a future state of super-fast expansion that pulls everything in the universe to bits.

2. There are good theoretical reasons to doubt the existence of phantom energy, but it is (as usual in this game) good to keep an open mind.

B. Another thing to keep in mind is that, when going from a single number (the vacuum energy) to an entire dynamical field (quintessence), we open the possibility of directly detecting the dark energy, just as we hope to directly detect the dark matter. That is, we can imagine that quintessence couples directly (although weakly) to ordinary matter.

    1. One effect of such a coupling would be a new force that stretched over macroscopic distances—a so-called *fifth force* (because the Higgs force is usually not counted). Experiments are currently underway to look for precisely such kinds of forces.

    2. It's good to keep in mind that, although so far our only evidence for dark matter and dark energy has been through gravity, we may someday be able to detect them in much more direct ways.

III. Quintessence has another virtue: It's observationally testable.

  A. Dark energy makes the universe accelerate because it has a persistent (nearly constant) energy density. But if the density is slowly varying, the rate of acceleration will be subtly different from the case where the density is truly constant, and that difference is, in principle, observable.

  B. The possibility that the dark energy density evolves with time is usually couched in terms of an *equation-of-state parameter*, often denoted w. This is the number that relates the pressure to the energy density.

    1. If the density is strictly constant, w is exactly -1, because the pressure is exactly the opposite of the energy density.

    2. If the density is slowly decreasing, w is slightly greater than -1 (that is, -0.9 or something like that); if the density is slowly increasing, w is slightly less than -1.

    3. This isn't an especially helpful way to think about things, but it's conventional among working cosmologists.

  C. To test the evolution of the energy density, we probe the behavior of the scale factor as a function of time.

1. In other words, we do something like the supernova projects did, although to much better accuracy.
2. A variety of new observational programs—using supernovae, lensing, and other cosmological probes—are now underway to do exactly this.

Recommended Reading:

Cole, *The Hole in the Universe*, chapter 8.

Nicolson, *The Dark Side of the Universe*, chapter 11.

Vilenkin, *Many Worlds in One*, chapter 14.

Questions to Consider:

1. The idea that the dark energy is a strictly constant vacuum energy, or cosmological constant, seems to fit all the data we have. Is there a good motivation for continuing to think of alternative models for what the dark energy might be?

2. Phantom energy could lead to a Big Rip in the future, but it's hard to know for sure. What would we need to be able to say with confidence in order to predict that certain things will happen in the distant future of the universe?

3. Can you think of different ways to measure the equation-of-state parameter of the dark energy? What about its value at earlier times in the history of the universe?

Lecture Eighteen—Transcript
Quintessence

Hopefully, by now, you can appreciate a little bit of the conundrum in which cosmologists find themselves. On the one hand, we have a theory that fits the data, the theory of dark energy being 70 percent of the universe, along with 25 percent dark matter and 5 percent ordinary matter. If that dark energy is vacuum energy, if it is strictly constant, 10^{-8} ergs in every cubic centimeter of space, unchanging as the universe expands, we can explain a whole bunch of observed phenomena all at once. We explain the flatness of space; we explain the fact that the universe is accelerating as a function of time. But, then, when we dig into that idea a little bit and start asking about how quantum field theory and our understanding of gravity would predict an energy density for empty space, we get a number, that prediction, that is bigger than what we actually observed by a factor of 10^{120}, a 1 followed by 120 zeros. This is the cosmological constant problem, the biggest issue right now in all of theoretical physics. What we used to believe before we thought that there really was dark energy in the universe is that the reason why the vacuum energy was small is because it is exactly zero, because there is some secret symmetry, some dynamical mechanism that we haven't yet discovered, which takes what we think should be large vacuum energy and squelches it all the way to zero.

Now we're in a slightly trickier situation. We want to squelch it most of the way, but not all the way there. One of the plausible scenarios is a two-step process, one that says the actual vacuum energy really is zero. It is still true that there is some unknown mechanism, which sets the vacuum energy exactly to zero. We haven't found it yet, but we'll still be looking. Then, we explain the dark energy by some separate mechanism, by some other form of stuff that acts dark energy-like. In other words, it is more or less the same in every different cubic centimeter of space and more or less constant through time, but not strictly so, something that can slowly change or gradually fluctuate from place to place. This would be a dynamical kind of dark energy, something that is temporary—something that can eventually go to zero. It might be in such a universe that, in the future, there won't be any dark energy anymore. One of the nice things about this possibility is that you can go test it. If the dark energy is not strictly constant, it is observationally distinguishable from the absolutely constant vacuum energy.

What we want to do is try to make some specific models, some theories of what a dynamical dark energy could be, and then test them against other things that we can observe in the universe. It's not hard to come up with models for dark energy that are dynamical. The simplest idea is one called *quintessence*, and it just fills the universe with a new kind of field. We know that the Standard Model already tells us that the universe is made of fields. There is a field for the electron; there is a field for the photon, the electromagnetic field, and so forth. If we want to add new physics, we're going to start adding new fields.

What are the properties that we want this particular new kind of field to have? For one thing, we want it to be a boson. If the field were a fermion, it would take up space so you could only have one new quintessence fermion in any one place. We want there to be a smooth distribution all over the place with a whole bunch of quintessence, relatively speaking. So, we want it to be a boson.

We also want it to not take out any preferred direction in the universe. If you think about some of the other fields that we know and love, like the electromagnetic field, when the electromagnetic field is turned on, the electric field is pointing somewhere. But, the universe looks the same in every direction, so we want the quintessence field to not pick out a direction in space. Physicists call such a field a scalar field, a field that just has a value everywhere. It's just a number. It is not a little arrow or a little vector pointing in some direction. The really crucial property the quintessence should have is that it should evolve very, very slowly. What this means is the observations are telling us that the amount of dark energy is more or less constant as the universe expands. Now, if you have a scalar field, you have a field that has energy, is a boson, does not pick out any preferred direction, there are scalar fields in the Standard Model. There is the Higgs field, but the Higgs field does not slowly evolve as the universe expands. It dramatically goes down to the bottom of its potential and just sits there. The Higgs field is not contributing any slowly changing dynamical contribution to the dark energy.

We previously offered an analogy to what a field is like. It's like a pendulum swinging back and forth, but there is a pendulum at every single point in space rocking back and forth. Then, the different physics of the different fields corresponds to what that pendulum couples to, what its amplitude is, and so forth. In other words, what we really want for quintessence is a pendulum that is almost stuck at some non-zero value and is going down very, very, very slowly, as if it's a pendulum stuck in very cold molasses and it cannot evolve very quickly.

We know how to do that. Physicists know how to write down theories that have bosonic fields, fields that can pile up, that move very, very slowly, that have energy densities that don't change very much as a function of time. However, for better or for worse, it changes the situation compared to if you only had vacuum energy. Vacuum energy is just a number. It's a value that, observationally, is something like 10^{-8} ergs per cubic centimeter, which tells you the minimum amount of energy in every single place in the universe if there is no stuff there. That number doesn't change; there is nothing that that number can do other than make the universe accelerate and otherwise have a gravitational field.

But, if you're adding an entirely new field to nature, that field has dynamics; it can interact. That's good because you can measure things about it; you can test it. It's interesting because those dynamics might have interesting new features, but it can also be problematic if those interactions have already been ruled out. Let me just mention one thing that you might want the quintessence field to do, one motivation for considering some dynamical feature of dark energy rather than just something that is absolutely constant. That motivation is called the *coincidence scandal*. The coincidence scandal is just the fact that in the current universe what we're claiming is that the dark energy is 70 percent of the total energy density. The matter density, both dark plus ordinary, is about 30 percent. So, 70 percent and 30 percent are not that different, and in the traditional cosmological evolution, these numbers change with respect to each other dramatically, as the universe expands. So, the energy density in dark energy is more or less constant from moment to moment. The energy density in matter is plummeting. As the universe expands, the number of particles per cubic centimeter is going down as the volume goes up. So, you have two numbers, the density of matter and the density of dark energy. They are changing dramatically with respect to each other. Yet, today, they are the same to within a factor of about two or three. The dark energy is about two or three times bigger than the matter density.

Why are we so fortunate to be living in just that moment of the history of the universe when the dark energy and the matter have comparable densities to each other? If you'd like, think about what life would have been like back at recombination. At the moment when the microwave background was being formed, the universe had a scale factor, a size, which was 1/1000th of its current size. At that moment, at recombination, the density of matter divided by the density of dark energy, would be one billion. There was a billion times more energy density in matter than in dark energy. That's a completely typical number. You're not surprised if you get a number like a billion. You're surprised to get a number like two because if a number changes from being a huge number in the past to being a really, really tiny number in the future, there is something special about now.

Ever since cosmologists realized that the earth is not the center of the solar system and Copernicus put us at the edge, we have been very, very wary of any theory that says there is something special about us, something special about us in space or as a function of time. But, the coincidence scandal says exactly that. If you have an absolutely constant vacuum energy, then it's very difficult to think what you could possibly do about the coincidence scandal. There is nothing that changes in that vacuum energy. There is no mechanism that makes it kick in at any particular time in the universe's history. It's set, by hand, in the very early universe and then you just get lucky later on.

I'll later discuss the possibility that the vacuum energy is very, very different in very different parts of the universe, parts we don't observe; therefore, you might get some explanation for the coincidence scandal within the context of a constant vacuum energy. But, if you don't believe that, if you believe that what you see in the universe is the kind of universe you get everywhere, didn't have any hope of explaining the coincidence scandal dynamically to some mechanism rather than just saying, well, we got lucky, you need to give the dark energy itself some kind of dynamics. That's what quintessence does. That's what the possibility of dynamical dark energy tries to do.

I'll make a confession. The confession is that even though it was a motivation for thinking about quintessence, attempts to actually explain the coincidence scandal by using quintessence haven't really worked. People have tried to invent theories where the energy density in quintessence didn't used to be constant. It used to decline very, very rapidly in the early universe, just like the density in ordinary matter of radiation. But, then, something changed relatively recently in the universe's history—for example, the formation of galaxies. Galaxies didn't exist when the universe was 1/1000th its current size. Maybe, when galaxies started to form, something changed in the dynamics of the quintessence to make it stop evolving. That's more of a hope than an idea right now. There is not really any theoretical models that make that happen, but that's the kind of thing that cosmologists are looking for. Right now, I can't, in good conscience, point to any models that actually make that happen, but that's not to say that tomorrow morning one won't appear on my desk. This is the kind of thing that we're trying to think of right now in the context of dynamical dark energy.

Another thing is that the future of the universe could be very different, if you believe in dynamical dark energy. The analogy I gave for quintessence is a pendulum that is going down very, very slowly because it's stuck in molasses. But, if the energy density of the dark energy is allowed to change, then we should be open-minded. We know almost nothing about the energy density of the dark energy or the fundamental physics behind it; therefore, if we can invent theories in which the dark energy is gradually fading away as a function of time, why not consider theories where the dark energy density is gradually growing as a function of time?

That's not to say it's easy to invent such theories, but once again, we can do it. Physicists know how to write down equations governing the behavior of a dark energy field for which the density would go up. So, not only the total amount of dark energy because the universe is expanding, but the amount of dark energy in every cubic centimeter would gradually be going up, as the universe expands. This is a very dramatic idea, which has been given the name of *phantom energy*, but it leads to a very dramatic consequence, which is why most people are interested in it. If you have an energy density in empty space that is growing and it continues to grow into the future, remember that what the dark energy does is to impart a constant impulse to the expansion of space. So, if the amount of dark energy is increasing and it's constantly giving an impulse to the expansion of space, the expansion rate will be increasing, the actual Hubble parameter will be going up. Space will be expanding faster and faster and you can reach a singularity in a finite amount of years into the future, a singularity in which everything is ripped apart. Individual galaxies, planets, and individual atoms are ripped apart by a huge amount of energy density in empty space. This has been called the *Big Rip*, to be a possible future evolution of the universe in contrast with the Big Bang that we had at the beginning—a singularity to cap off the end of the universe just like there was a singularity that started it.

All of this is completely hypothetical. There is no evidence that we have right now that something like the Big Rip would be happening. In fact, there are important physics worries about these models. If you can have energy grow, what that means is that if you look at the excitations of this field, if you look at the particles that you would see if you actually observed the quintessence field directly, what you would see are particles with negative energy. So, in otherwise empty space, you could spontaneously create some positive energy particles and some negative energy particles, without violating conservation of energy. That's never been seen. It would mean that lighter particles could decay into heavier particles by emitting particles with negative energy. That's never been seen.

So, the possibility of phantom energy is not a leading candidate. Most physicists are kind of either scared or appalled by the idea. But, it's the kind of thing that we're driven to think about because we know so little about the underlying physics of dark energy. It's the kind of thing to keep in mind about what the possibilities include.

Let's go a little bit more down to earth and ask, if there is dynamical dark energy, what does that get us? Who cares? Can we somehow do something with it in terms of observational constraints on what it is doing? The answer, asked in that way, is of course, yes. Once we have a field, not just a number that is the same everywhere but a field that can vary from place to place and has dynamics of its own, that field can interact. The dark energy field, the quintessence field, can interact with ordinary matter, with dark matter, or both. A whole bunch of opportunities open up.

How would you notice if the quintessence field was interacting with ordinary matter? Remember, the quintessence field is a number that is slowly changing everywhere in space. It's slowly rolling down some potential field, so that its energy density is gradually evolving as the universe expands. Two things are evolving: One is the value of the field; the other is the amount of energy contained within that field. Every given theory is going to tell you for a given value of the field how much energy is contained in it. But, because the field itself is evolving, its interactions with all of the rest of nature mean that there'll be hidden effects on the rest of the particles, for example, in the Standard Model. For example, you would expect, if you were in a background field that was slowly changing, for things like the mass of the electron and the charge of the electron to be gradually changing as this field evolved. It's not necessary. You can certainly invent models where that doesn't happen, but the default assumption is that would happen.

In other words, you can look for quintessence directly in the behavior of ordinary matter by asking if the so-called constants of nature are truly constant or if they are changing. As it turns out, we have a lot of data that tell us that the constants of nature are the same now as they were in the past. If you go all the way back to Big Bang nucleosynthesis, a minute after the Big Bang, we have made very precise predictions for the abundance of helium, lithium, and deuterium on the basis of our current knowledge of nuclear and atomic physics.

If something were different, if the mass of the proton were different during the Big Bang nucleosynthesis, you would predict in principle very different abundances for the light elements. The fact that you get the right answer in conventional nucleosynthesis tells us that the constants of nature at that very, very early time were more or less the same as they are now. There are similar phenomena that we don't have time to go into in great detail. There is something called the Oklo natural reactor, which was a naturally forming formation in Gabon in West Africa where there was a self-sustaining nuclear chain reaction. Billions of years before Enrico Fermi put up the first man-made chain reaction outside of the University of Chicago, nature did it itself, turning heavy uranium into lighter elements, and we can go there today and measure the reaction products. We find results that are consistent with the hypothesis, that the constants of nature back then are the same to within 1 part in 10 million as they are today. So, the data are telling us, as far as we can tell, the constants of nature two billion years ago are the same values as they are today.

Quintessence would tell us that it would be very, very plausible for them to have changed. That's not a rock steady limit. It is not absolutely ruling out the idea that there is quintessence, but it's putting some pressure on it. If there had been quintessence, maybe we should have seen it already in something like the decay of the constants of nature.

The other possibility, if quintessence interacts with ordinary matter, is new forces. Remember, bosonic fields give rise to forces—the photon, the graviton, the gluon, etc. Furthermore, there is a rule that says the range of the force, the spatial distance over which the force can stretch, depends on the mass of the boson that is carrying that force. The forces that we know in nature that go over long range are the gravitational force and the electromagnetic force. These are the ones that we can see in our macroscopic everyday lives. The reason why these stretch over long distances is because the bosons that carry them, the graviton and the photon, are massless. If the bosons are massive, they give rise to very short range forces, and that kind of makes sense. It just takes energy for the boson to stretch over large distances. If it has a large mass, it's not going to stretch very far.

What about quintessence? Quintessence is a boson that has a very small mass, very close to zero. If the mass were large, it would have fallen to the bottom of its energy already and it would not be evolving. By hypothesis, the quintessence does not have a large mass; therefore, it should give rise to a long-range force. This would be what particle physicists call a *fifth force* of nature because we already have gravity and electromagnetism, the strong and weak nuclear forces. This would be a new force and a new force that stretches over macroscopic distances. It would be kind of like gravity, a weak new force, except that unlike gravity, it would not be universal. The secret to gravity is that everything falls in the same way. Every object, regardless of what it is made of, feels exactly the same gravitational force.

If quintessence exists, it gives rise to a new force, but one that affects different objects differently. So, we're actually looking for exactly that. Experiments are going on to measure the acceleration of little balls made of different substances in the direction of the sun. We are able to measure the force due to gravity of balls here on earth caused by the sun by looking at how much they move in that direction at some time of day versus another time of day. We can do this again and again with different kinds of substances. What we find is there is no evidence for a new force of nature that depends on the composition of the different kinds of stuff you're dealing with. Once again, it is not saying that quintessence doesn't exist, but there is a chance for us to find it in fifth force experiments and we haven't found it. The supposition needs to be that if quintessence is there, somehow it is hiding from us. Somehow, there is some symmetry or some dynamical mechanism that is preventing us from directly detecting the quintessence in any obvious way.

However, I don't want to be too pessimistic about things. I want to emphasize the fact that we're just beginning to measure the physics of the dark sector. We're proposing that 95 percent of the universe is made of stuff that we haven't directly seen—dark matter is 25 percent and dark energy is 70 percent. The model we have right now that fits the data is a very minimal, vanilla kind of model. It says that the dark energy is strictly constant and doesn't interact with either ordinary matter or dark matter except through gravity. The dark matter is completely non-interacting, 25 percent, and doesn't interact with dark energy or with ordinary matter except through gravity. That is a model that fits the data.

But, again, we're nowhere close to having completely figured out the physics of dark matter and dark energy. So, we could be seeing something much more dramatic. A hundred years from now, we might have a whole, rich phenomenology of the different kinds of interactions that characterize the physics of dark matter interacting with ordinary matter, dark energy interacting with dark matter, and ordinary matter interacting with dark energy. We're just at the beginning of thinking about these things, so we don't know what's going to come until we go out there and look.

I should mention that quintessence, the idea of a single field slowly rolling down, is not the only way we could imagine getting dynamical dark energy. It is by far the leading candidate. It's very easy to write it down. It's very easy to come up with specific models that fit the data, but there are alternatives. For example, there is the idea of tangled strings in the universe. We will talk about *superstrings*, very, very tiny strings whose vibrations look like elementary particles. There are also *cosmic strings*, a very, very different idea, strings that stretch across the observed universe. We haven't seen any of these things yet, but if the cosmic strings are relatively light, we would not have seen them.

If the cosmic strings have the property that when two strings bump into each other they get tangled, the total energy density in cosmic strings can evolve very slowly, if at all. In other words, they can be something like dark energy. It turns out that when you run the numbers, the actual energy density even in tangled cosmic strings seems to go down too quickly to be the dark energy. But, again, maybe that's just because there is something we're missing in the models. It's something to keep in mind.

The other possibility is something I like very much because I helped invent it. It's the possibility of variable mass particles. What if you really, really wanted to believe that the dark energy, just like the dark matter, was made of particles? The real problem with believing that is that particles, slowly moving particles, have an energy $E = mc^2$ that doesn't change as a function of time. Since the energy per particle doesn't change, as the universe expands, the energy density goes down and that's not what dark energy does. But, imagine that you had particles whose masses went up as the universe expanded. In other words, the energy per particle was still $E = mc^2$, but the mass was going up just like the number density was going down. Then, you could have the total energy density in this kind of stuff act like dark energy. These variable mass particles, or vamps as we called them, could be a kind of particle that didn't have its energy density go away as the universe expanded.

The bad news is, in the details, it doesn't quite work. You might want to ask, how should the mass change? What is governing the value of the mass of each particle? What you would do is you invent a scalar field. In that field, you would invent a field that governs the mass of the particles, and that would basically act just like quintessence. It turns out that the idea of very well-masked particles is not really a separate idea from quintessence. It's a way in which you can have the dark matter particles interact with the dark energy field, which is something interesting to think about. So far, it is not something that is pushed upon us by the data, but it might be there so we're still looking.

Let me finish up by talking about how we would know, how we could actually go about testing this idea that the dark energy is dynamical and not strictly constant. The way that we found the dark energy in the first place was to look at the acceleration of the universe. If the dark energy weren't dark energy, if it were just matter, you would have a decelerating universe as all the particles in the universe pulled on all the other particles, and the universe would expand but ever more slowly as time went on. When the dark energy kicks in, it's a constant energy density that provides an impulse to the universe and we see acceleration. So, if you imagine that the dark energy is slightly dynamical, that the energy density is not strictly constant, but slowly changing, then you still get acceleration, but you get a slightly different rate of acceleration than you would with vacuum energy, a strictly constant amount of energy density in an empty space.

So, cosmologists have invented a number to characterize how much the dark energy density evolves, as the universe expands. For weird, historical reasons, they call this number the *equation-of-state parameter* and they label it w. Remember when we talked about the way in which dark energy evolves and the reason why it makes the universe accelerate, we said that one way to think about the fact that the dark energy makes the universe accelerate was to say that it has a negative pressure. The thing that makes the space expand is the energy density plus three times the pressure. So, ordinary dark energy in the sense of vacuum energy, something that is strictly constant, has a pressure that is exactly equal but opposite to its energy density. Pressure equals -1 times the energy density. The idea of w, the equation-of-state parameter, is just to replace that -1 by an arbitrary number called w. If w is very close to -1, then the dark energy density is almost not evolving. If W is a little bit greater than -1, which, because -1 is a negative number means something like -0.8 or -0.9, that means that the energy density will slowly be declining.

That's the obvious guess. If you believe in quintessence, it's that the equation-of-state parameter should be a little bit greater than -1because the energy density in dark energy should be slowly going down, as the quintessence field evolves. But, it could be slightly going up. The dark energy density could be going up if you get phantom energy, so it could be that w, the equation-of-state parameter, is less than -1. It could -1.1 or -1.2. How would you know? You look at the data, the same kinds of data we used to discover the acceleration of the universe, and you fit it to a different kind of model. What we used to do was to fit the data to a model in which you had both matter and dark energy. The dark energy was constant, but the total sum of the two was arbitrary. Then, you asked, what fit the data? The thing that fits the data is something where the total amount of energy is the critical density and space is flat. So, you get a two-parameter family of possibilities—how much dark matter, how much dark energy?

You can replace that with a different two-parameter family of possibilities by assuming that space is flat. Assume that the total amount of dark energy plus the total amount of matter equals the critical density. Then, the two parameters you now have are the total amount of matter, which determines the total amount of dark energy, and the equation-of-state parameter, and w, which tells you how fast the dark energy evolves.

If that's true and you plug in the data, you get limits on what w is. Right now, that limit is something like w is -1 +/- 0.3. So, to a good confidence level, w is somewhere between -0.7 and -1.3. On the one hand, that's telling us that it's close to -1. The dark energy density is not evolving very appreciably as a function of time. On the other hand, it's telling us that there is room for improvement. Certainly, if the equation-of-state parameter was -1.1, we would not have noticed it yet, or if it were -0.9. So, we want to do better. Perhaps the biggest single experimental project in modern cosmology is trying to measure the equation-of-state parameter to higher precision.

We'll talk in Lecture 23 about a suite of new experiments. They are trying to pin down the equation-of-state parameter to +/- 0.05, to 5 percent, instead of +/- 0.3. To do that will require a lot more data, perhaps going to space, certainly building things here on earth. We will do it, though, and it's very important to do it, because the kind of physics that you invoke to explain constant energy density in empty space versus variable energy density is completely different. We may never know which one is right. We may get really unlucky. If the true equation-of-state of the dark energy in the real world is -0.99, it is hard to imagine how we will ever tell that it's not exactly -

1. But, in the meantime, we can hope that we're a little bit luckier than that, that we'll keep measuring it better and better, and pretty soon, if it comes closer and closer to being -1, we'll be able to say, yes, indeed, the dark energy that is 70 percent of the universe is the vacuum energy of empty space itself.

Lecture Nineteen
Was Einstein Right?

Scope:

We haven't detected dark matter and dark energy *directly*—at least not yet. We have only inferred their existence through the gravitational fields they cause. Is it possible that our understanding of gravity itself is incomplete, and that, instead of invoking new dark stuff, we should modify Einstein's general relativity? It is entirely possible, and several specific scenarios have been proposed. But we also have good experimental constraints on the behavior of gravity. It turns out to be very difficult to modify gravity successfully without running afoul of such constraints; attempts to do so are an active area of current research.

Outline

I. We believe in the dark sector because we infer its existence, not because we detect it directly.

    A. As we have emphasized, there is considerable observational evidence in favor of dark matter and dark energy.

        1. However, all that evidence proceeds in a very specific way: We measure gravitational fields (the curvature of spacetime), deduce what kinds of sources are necessary to produce such fields, and find that ordinary matter is insufficient.

        2. The deduction necessarily involves our understanding of gravity in the form of Einstein's equation of general relativity.

        3. But what if general relativity is not right? More specifically, what if it is an excellent approximation to reality in the Solar System (where we have tested it to very high precision, and it passes with flying colors) but increasingly deviates from the right answer on the larger length scales characteristic of cosmology?

    B. It's worth emphasizing that there's nothing sacrilegious about contemplating alternatives to Einstein's theory.

1. General relativity is a cornerstone of modern physics, but nobody believes that it is absolutely the last word on gravity.

2. If nothing else, general relativity is expected to break down on very small scales, when quantum gravity becomes important (as we'll consider in Lecture Twenty-One).

3. It's perfectly acceptable—if something of a long shot—to wonder whether it might break down on very large scales, as well.

C. Planets in the Solar System provide an ambiguous historical lesson.

1. In the 19th century, French mathematician Urbain LeVerrier suggested the existence of an undiscovered celestial body—the dark matter of its day—to help explain the orbit of Uranus.

2. Sure enough, in 1846, astronomers discovered the planet Neptune. Encouraged, LeVerrier tried again, proposing a new inner planet to help explain the discrepant orbit of Mercury.

3. But this planet, labeled Vulcan, was never found; as we now know, Mercury's orbit is explained by improving our theory of gravity from Newton's inverse-square law to Einstein's general relativity.

D. We can ask the question separately for dark matter and dark energy.

1. Evidence for dark matter comes from the dynamics of galaxies and clusters; generally, there is a stronger attractive force due to gravity than we would ordinarily expect.

2. Dark energy, meanwhile, is deduced from the global behavior of the universe: the accelerated expansion and the flat geometry of space.

3. Dark energy, in contrast to dark matter, appears to be repulsive, causing galaxies to accelerate away from each other.

4. The best possible theory of modified gravity, of course, would account for dark matter and dark energy in one

fell swoop; it would, nevertheless, still be interesting to do away with one of them, even if we had to keep the other.

II. Doing away with dark matter is a difficult proposition.

    A. The most celebrated attempt to replace dark matter with modified gravity is credited to Mordehai (Moti) Milgrom, who invented a theory called *modified Newtonian dynamics* (MOND). Milgrom looked at the rotation curves of spiral galaxies and noticed something interesting.

        1. In the central regions, no dark matter was needed; the visible matter was able to account for the observed dynamics.

        2. Dark matter was only necessary past a certain point, which is to be expected.

        3. The unexpected thing was that, for all sorts of different galaxies, the acceleration due to gravity at the point where dark matter became important was the same.

        4. Milgrom suggested that this was not an accident, that there is no such thing as dark matter, but instead, that the character of gravity (how fast the gravitational force falls off with distance) changes when the acceleration becomes very low.

    B. One big strike against Milgrom's theory was that it was incomplete.

        1. General relativity makes specific predictions for galaxies but also for many other phenomena—the Solar System, lensing, gravitational waves, and cosmology, just to name a few.

        2. MOND didn't, for a long time, have anything to say about such things.

        3. However, in 2004, Jacob Bekenstein finally invented a theory that could answer all these kinds of issues; in fact, it did pretty well in confronting the data.

    C. The real problem for MOND in its current form comes from clusters of galaxies. Clusters are often in the regime where the MOND rules should apply, but they don't seem to fit what is observed.

1. Even MOND's boosters now concede that there is only one way to make the theory fit the cluster data: dark matter!

2. The idea is that you could have hot dark matter, such as neutrinos, in the clusters; the hot particles would stream out of galaxies, which are explained by modified gravity, but remain stuck in clusters. This idea is currently compatible with the data, but it doesn't really add up to a compelling picture.

III. Doing away with dark energy is more plausible, although that still faces difficulties.

A. Compared to dark matter, we know rather little about dark energy. At a certain point, late in the universe's history, it begins to accelerate at a certain rate—we don't know much more than that. There is correspondingly more room to try to invent a theory of modified gravity that can fit the data even without dark energy.

B. However, modifying gravity turns out to be very difficult in practice. It's easy (as these things go) to modify gravity to make the universe accelerate; however, it's very difficult to invent such a theory that remains compatible with what we know of gravity from other experiments.

C. Nevertheless, we plunge forward.

1. The good news is that the acceleration of the universe isn't the only cosmological manifestation of gravity; there is also the evolution of structure, from small ripples in the CMB to galaxies and clusters.

2. Modifying gravity to make the universe accelerate will generically alter the predicted pattern of structure evolution.

3. The next generation of observations, therefore, will place special emphasis on comparing the implications of the expansion history of the universe to the history of structure growth.

Recommended Reading:

The ideas discussed in this lecture are sufficiently speculative and new that no good book discusses them in any detail. For some general ideas about gravity and how it might be modified, see the following references.

Greene, *The Fabric of the Cosmos*, chapters 13–14.

Randall, *Warped Passages*, chapters 18–22.

Will, *Was Einstein Right?*

Questions to Consider:

1. Einstein was a smart guy, and his general theory of relativity has already passed a great number of observational tests. Does it make sense to keep inventing new theories rather than stick with the one that works?

2. Do you think we'll ever be able to know for sure whether there is dark energy or, instead, whether gravity is modified? Likewise for dark matter?

Lecture Nineteen—Transcript
Was Einstein Right?

The ideas of dark matter and dark energy obviously play an important role in modern cosmology. They've made everything fit together in the sense that a whole bunch of data of a whole bunch of different kinds of phenomena suddenly makes sense, if you believe that only 5 percent of the universe is ordinary matter, 25 percent is dark matter, and 70 percent is dark energy. It is an impressive amount of data and it's not just the same kind of phenomena over and over again. It's a wide variety of different things that we're looking at that convince us that 95 percent of the universe is this dark stuff, this dark sector, so physicists naturally are very excited about what this stuff is, what its properties are. People are buying series of lectures trying to understand where this stuff comes from and what it might be.

However, we have to keep in mind that every single piece of evidence that we have that points in the direction that there are such things as dark matter and dark energy come from measuring gravitational fields. We do not find dark matter and dark energy yet *directly* in our laboratory. We hope to do that and we're trying to do that, but so far it's all been about inferring the existence of dark matter and dark energy on the basis of the gravitational fields we observe in the universe, on the basis of the way in which space and time are curved. We attribute that curvature to stuff. We can attribute it to a bunch of cold massive particles called dark matter and a smoothly distributed nearly constant kind of stuff in empty space called dark energy.

However, logically speaking, it's certainly possible that our inference is incorrect because the theory of gravity that we're using to relate observed gravitational fields to the stuff that causes them is wrong. That theory of gravity, of course, is Einstein's theory of general relativity. In this lecture, we'll dig into the possibility that general relativity is not right. Certainly, general relativity is pretty close to being right, both in the solar system, where we've tested it very well, and in other more local things in our galaxy like the Binary Pulsar, a pair of two neutron stars that is changing its rate of rotation due to admitting gravitational waves just like Einstein predicted. In other words, we have a lot of empirical data saying that general relativity is a pretty good theory on scales of several astronomical units, one light-year across or less.

But, now, in cosmology, we're looking at the whole universe. We're looking at many millions of light-years across, many billions of light-years across; it is certainly possible that gravity acts differently on length scales of billions of light-years than it does on a scale of one light-year. Can we do better than Einstein? Can we come up with a better theory than he did that will explain how gravity works on large scales and fit the data without invoking dark matter or dark energy?

I should say that Einstein, of course, is a famous physicist and he's a very smart guy. But, even though his theory is very good, there is nothing sacred about it; therefore, there is nothing sacrilegious about questioning general relativity. There is a feeling that some people have, that we establishment cosmologists have, as our duty, to defend Einstein's honor against attacks from talented but quirky outsiders. Nothing could be further from the truth. Any working theoretical physicist would like nothing more than to be the one who came up with a better theory of gravity than Einstein. That's why we all got into the game, so that we could have a better understanding of nature, something that goes beyond what we already have.

Furthermore, we all believe that general relativity is not the final true answer. We all expect that general relativity needs to be modified in some way, if only because general relativity as we currently understand it is completely inconsistent with our understanding of quantum mechanics. That's the problem with quantum gravity, which we'll be getting to in a later lecture, the string theory has been purposed to resolve it, but it tells us that general relativity is at best an approximation. It can't be the final correct answer because the world is not classical and general relativity is. Therefore, maybe general relativity is also wrong on larger scales. That's not what we would expect. Usually in physics if we understand something at a different physical distance, we will also understand it at larger distances. New things will kick in at smaller distances where energies become larger and quantum mechanics becomes important. But, logically speaking, it is a possibility, so we're all very, very interested in wondering whether or not our current inference about dark matter and dark energy could somehow be traced to a misunderstanding of how gravity works.

Before we go into details, it's worth telling this very amusing parable, the story about how people have already tried this game and how we get ambiguous results from looking at the historical record. Most of what we know about gravity, both even today as well as in the past, comes from the dynamics of stuff in the solar system—the different satellites around the planets and the planets moving around the sun. Everything moves in a gravitational field. This is how Kepler was able to find out that the planets move in ellipses rather than in perfect circles. From Kepler's understanding of planetary motions, Isaac Newton was able to derive his inverse-square law for gravity—the idea of the force, the gravitational force between two objects, decreased inversely as the distance squared.

However, this theory of Isaac Newton's, even though it fit the data of the planet very, very well, didn't always fit perfectly. The first time anyone discovered a new planet since ancient times was the discovery of Uranus in 1781. It was discovered with a telescope, a brand new planet out there, once again evidence that the heavens were not fixed and given to us ahead of time. We had to go out there and look to see what would be there. People were very excited about this. They took many different observations of the motions of this new planet and they tried to ask whether or not Uranus fit into what we understood about the solar system.

Interestingly, the motions of that new planet were not precisely those that you would have expected, given the laws of gravity, as they had been handed down by Isaac Newton. What do you do when there is an object in the sky whose dynamics are not exactly what you would of predicted on the basis of the gravitational fields you think are there? You invent dark matter; you invent some new stuff that you haven't directly seen, whose gravitational field is influencing the motion of the things you do see. The French mathematician Urbain LeVerrier invented a new planet, which we eventually called Neptune. He predicted exactly where Neptune would have to be located in the sky, in order for it to be pushing around the orbit of Uranus in the right way. When he finally convinced people to go look for it, they found it to within one degree of where he predicted that it was in the sky. This is the first example in history of where a dark matter hypothesis was presented and turned out to be right. Of course, once you see it, it's no longer dark matter, but we're hoping to do exactly the same thing with real dark matter and real dark energy.

Then, it was realized within LeVerrier's lifetime, that there was also a problem with the inner planets. The motion of Mercury, in particular, was not precisely what Isaac Newton said it should be. Mercury moves in an elliptical orbit and that ellipse was known to precess, to change its orientation gradually as a function of time. Part of this precession was just due to the fact that the solar system is not a pristine place; there is more to it than just the sun and Mercury. The sun is rotating itself, which gives it a different kind of gravitational field, and of course, there are the other planets. So, LeVerrier, fresh off of his success of predicting dark matter and discovering Neptune, did the same trick again. He proposed that there is dark matter in the inner solar system, in the version of a new planet called Vulcan. Vulcan would be another minor planet in the solar system, inside the orbit of Mercury. LeVerrier predicted exactly where it should be, and lo and behold, it was discovered. In fact, it was discovered more the once. All of these times it was discovered turned out not to be right because, of course, there is no planet there.

In the case of Mercury, the reason it is not moving in the way that Isaac Newton predicted that it should is because Isaac Newton wasn't right, because gravity doesn't quite work the way that Newton said it does. There are slight deviations; those deviations were finally understood when Einstein came along and made his theory of general relativity.

After Einstein invented general relativity, one of the first things he did was to check what it would predict for the orbit of Mercury. He found that it exactly accounted for the discrepancy between what was observed and what was predicted. This, Einstein said, was the happiest moment of his life when he realized, before anyone else in the world did, what the correct explanation was for the motion of Mercury.

What is the lesson from our little story? We can see that things move through the heavens; we can use them to predict the existence of stuff, based on our understanding of gravity. Sometimes that prediction turns out to be right. Sometimes we do understand what gravity is doing and we successfully infer the existence of new stuff. Sometimes it's because gravity is the culprit, so we can try to do exactly the same thing now; we can try to use a new theory of gravity to try to explain away dark matter and dark energy. As one final bit of caution, we should notice that in the solar system, the place were the dark matter hypothesis worked, was in the outer-solar system. The place where we didn't understand gravity was at shorter distances, at smaller length scales, so the fact that we understand gravity right now very, very well in the solar system suggests that we probably understand it in the galaxy and in cosmology. But, once again, because the very discovery of dark matter and dark energy has been such a surprise, we're trying to keep an open mind to see what will happen.

The evidence we have for dark matter and dark energy takes very different forms for those two different things. We believe that dark matter is something local, something that falls into places where there is ordinary matter, into galaxies and into clusters, and increases the strength of the gravitational field. It's a fairly simple idea to try to get rid of dark matter with a modified theory of gravity. The fact that you have a galaxy or a cluster, and you measure its mass by looking at the motions of things near the edge, and you find too much, could potentially be explained by the simple hypothesis that the force due to gravity is stronger at large distances than you thought it would be. In other words, it falls off more slowly than one over R squared that Newton would have you think. That's the kind of thing you would need to make work if you want to replace dark matter with modified gravity.

Dark energy, on the other hand, is smooth and global. It does not collect into galaxies and clusters; it is all over the place. Furthermore, it does not pull things together more strongly like dark matter does; it is pushing the universe apart. The first manifestation of dark energy was that the universe is accelerating at a faster and faster rate. So, to get rid of dark energy, rather than dark matter, what you need to do is to find something that makes the universe accelerate and something that is not associated with any given kind of source—something that is all over the place, not collected in the area where there is ordinary matter. So, the attempts to do away with dark matter and with dark energy by modifying gravity seem to have a different character. Maybe they are the same modification or maybe they are two different modifications. Maybe there is dark matter, but not dark energy, or vice versa. All of these possibilities are there on the table.

Let's think in more detail about the dark matter case. What do we actually know? What is our evidence for dark matter? The original evidence, of course, came from clusters and from individual galaxies. We would look at the dynamics of these systems. We would look at stuff moving around galaxies, for example, and use that stuff to infer the total amount of mass. We would look at the motions of galaxies and use the velocities at which they are moving to infer the total amount of mass in a cluster of galaxies. These days, we have slightly more sophisticated forms of evidence that are slightly less direct. We have the cosmic microwave background, the different splotches in the microwave background respond to the existence of dark matter, and that's more evidence that there is more in the universe than just ordinary matter.

We have the evolution of structure in the universe. For that matter, we have the fact that the universe is spatially flat, which relies to some extent on the existence of dark matter. It's kind of remarkable that the one hypothesis of dark matter particles can fit all of those different phenomena. It would be really difficult to imagine coming up with a new theory of gravity that did just as well. We can start with a simple starting point. Forgetting about trying to explain everything at once, let's just take one phenomenon and try to explain that.

This was done in 1984 by Mordehai Milgrom, who said let's just look at the rotation curves of spiral galaxies. Remember, the rotation curves of spiral galaxies were measured by Vera Rubin in the 1970s, and were the first really quantitative evidence that most of the stuff in the universe was dark matter, not ordinary matter. She measured the velocity of different kinds of stuff, as you went from the center of a galaxy to the far edges. If what you saw in the galaxy was what there was, that velocity would fall of as you got further and further away. What Rubin found was that that didn't happen. The velocity stayed nearly constant, evidence that you need dark matter in the outskirts of the galaxy to make the rotation curves make sense in the context of Newtonian gravity.

Milgrom noticed a very interesting fact about the universe. Regardless of whether or not he's correct about modifying gravity, this fact remains true. The fact is that if you look at different galaxies, measure their rotation curves, and look at where you need dark matter and where you don't, you find that in the centers of galaxies, you don't need to imagine that there is dark matter. Maybe it's there, but the ordinary matter is enough to explain the motions that you observe. It's only in the outskirts that the effect of dark matter builds up enough that you absolutely need to imagine that it's there to explain the data.

The radius of the region where ordinary matter is enough is different for every galaxy and everyone knew that. But, what Milgrom noticed was, if you calculate, just using the laws of Isaac Newton, the acceleration due to gravity at the point where ordinary matter is no longer sufficient, the point where you need to invoke dark matter to explain the motions, that acceleration due to gravity was the same in every galaxy. There was some crossover value of the acceleration, such that when the acceleration due to gravity was larger than that value, you didn't need dark matter. When it was smaller, you did. What Milgrom said is that maybe the point is that there isn't any dark matter; maybe there is a new theory of gravity. Maybe the theory of gravity that is correct has the property that if the acceleration is greater than a certain value, it's a one over R squared inverse-square law just like Newton said, but at smaller values of the acceleration, gravity has a force that is stronger than that. It falls off more gradually like one over the distance, rather than the distance squared.

Whether or not that's true, it fits the data very well, not only for the galaxies that Milgrom knew about when he was making this hypothesis, but for new kinds of galaxies that were discovered after he made it. For individual galaxies, Milgrom's rule of thumb, that you don't need dark matter, Newtonian gravity works in the insides, and there is a very specific radius outside of which you can put the data by modifying gravity, fits over and over again. This a remarkable fact about the universe. Maybe it's because there is modified gravity; maybe it's because cold dark matter arranges itself, so that that will be the case. If the latter possibility is right, it's a challenge for the theory of cold dark matter to predict why this observed fact found by Milgrom seems to be so true in the real world.

The problem with Milgrom's theory, which is called *modified Newtonian dynamics* or MOND, is that it's not a theory. It's a suggestion for one very specific circumstance, for the rotation curves of galaxies. But, we know there is more to the universe than galaxies. There are clusters of galaxies. There is the expansion of the whole universe. There are more phenomena than rotation. There are galactic gravitational lenses, for example. How can you make predictions for all of these observed phenomena from Milgrom's theory? The answer is, you can't. You need to take that suggestion, that the law of the fall off due to the gravitational force is different than Newton's, and imbed it in a more comprehensive framework from which you can make better predictions.

Even before you do that, the one thing that you can do is try to go from galaxies to clusters of galaxies. What we found over and over again is that it doesn't work. Milgrom's modified rule works very, very well for individual galaxies, but when you start including clusters of galaxies, you could make a prediction and the prediction did not come true. People tried and tried to understand clusters of galaxies better and better within Milgrom's framework and they never quite succeeded. Eventually, they more or less gave up on the idea that you could explain clusters of galaxies purely by modifying gravity, at least with Milgrom's idea. The people who are now proponents of MOND, proponents of Milgrom's idea, believe that in order to explain clusters of galaxies, there has to be some dark matter there. However, they say that the dark matter that is in clusters of galaxies might be different than the kinds of dark matter you imagine in conventional concordance cosmology. For example, the dark matter that is in clusters of galaxies could be neutrinos. In ordinary cosmology, we don't think the dark matter can be neutrinos because neutrinos are hot dark matter. They move very quickly; they don't settle into galaxies and make them form. In a universe dominated by neutrinos, structure would be very, very featureless. But, they do move slowly enough that they can be captured by clusters of galaxies. So, basically, people who believe in Milgrom's theory have retreated into a position where clusters of galaxies are explained by dark matter, but individual galaxies are explained, without dark matter, by a modified theory of gravity.

Still, it wasn't until 2004, 20 years after Milgrom's original theory, that Jacob Bekenstein finally managed to come up with a full theory that reduced to Milgrom's idea in the appropriate circumstances. You'll not be surprised to learn that Bekenstein's theory was very complicated. It's not easy to come up with a modification of gravity that fits all the data. Bekenstein's theory invented new fields. That's what you do whenever you invent a new theory of anything; you add new fields and see what they do. He added a bunch of new fields. He found that he could predict what Milgrom predicted for the cases of individual galaxies, but you could also fit other cosmological data. Remarkably, Bekenstein's theory fit things like the microwave background and the acceleration of the universe, if you not only put in the neutrinos you needed to explain clusters, but also you put in dark energy. On the one hand, you finally had a model of modified gravity in which individual galaxies did not have important amounts of dark matter in them. But, other than that, it was a very good conventional model. Clusters of galaxies and the whole universe are defined by dark matter and dark energy. So, even though that kind of model fits the data, the people who were originally enthusiastic about it have lost a certain amount of motivation. Originally, the hope was you were getting rid of dark matter entirely. Now you're just getting rid of non-Standard Model dark matter hoping that it could be neutrinos, but the motivation for doing that is a little bit less. But, it still could be true.

Let me back up a little bit from the notion of Milgrom's particular proposal to the general notion of doing away with dark matter in galaxies. Let me mention my belief that you absolutely need some kind of dark matter in the universe. It is impossible in principle to think of a theory in this day and age, which will completely do away with dark matter. There are two pieces of evidence for that. One is, of course, the Bullet Cluster; we talked about the Bullet Cluster when we talked about the original evidence for dark matter. Here you have two clusters of galaxies that pass through each other, the hot gas that is most of the ordinary matter got stuck in between, and the galaxies and dark matter went right on through. When you look for the gravitational fields of these clusters using gravitational lensing, you find that most of the gravity is being caused by the dark matter, not by the ordinary matter that got stuck in the middle. If you're going to modify gravity in a way to get rid of dark matter, typically what you will do is you will modify the strength of gravity as a function of distance. But, it is very hard to modify the direction of gravity. If you are trying to come up with a model in which there is no dark matter, the only source for gravity is the ordinary matter. The only plausible direction for which the gravitational force can point in is toward the ordinary matter. But, the Bullet Cluster is a very clear-cut example where we have a gravitational field that is pointing in a direction other than where the ordinary matter is. The Bullet Cluster makes perfect sense if there is dark matter. It is very hard to understand if there is no dark matter and gravity is modified.

Similarly, you have the cosmic microwave background. We told the story about the oscillations of over-dense regions in the early universe, and how sometimes they were in phase with the dark matter, sometimes they were out of phase, and that is precisely what you observe in the cosmic microwave background. If there is no dark matter, if there is just modified gravity, then there isn't anything to be in phase with or out of phase with. It becomes correspondingly much, much harder to explain what we see in the cosmic microwave background. These are both very general arguments; they don't attack any specific proposal, but what they say is that we have very good evidence that there isn't any specific proposal that you can possibly come up with to do away with dark matter and replace it by modified gravity. You never know; it's not a theorem that is absolutely rigorous, but it's pretty good evidence that dark matter really does exist.

What about dark energy? Dark energy is very different than dark matter. Among other things, it's much simpler or at least we know much less about it. Dark matter clusters together; it has dynamical properties. Dark energy might be dynamical, but the only thing that we know is that it's a nearly constant energy density, enough to make up 70 percent of the critical density of the universe. What you want to do to modify gravity, to get rid of dark energy—in other words, imagine that the dark energy is zero, but there is some modification of Einstein's equation that makes the universe accelerate—is ultimately to modify the Friedmann equation. The Friedmann equation is the one that tells us how the expansion of the universe responds to energy density and it tells us that if the energy density goes to zero, the expansion rate goes to zero along with it.

If you want to do away with dark energy, you want to modify the Friedmann equation in such a way that, as the energy density goes to zero, the expansion rate stops going to zero. Either it goes to zero much more slowly than the energy density does or it actually gets stuck at some finite value. The important thing to realize here is that there are constraints on what you can do. Just like with dark matter, there are plenty of experimental constraints. For dark energy, where we want to modify the Friedmann equation, there are some things we know about the Friedmann equation. For example, we know it works very, very well in the early universe. The Friedmann equation is tested to very good precision by both primordial nucleosynthesis and by the cosmic microwave background. We have two sets of phenomena, which if you try to explain them on the basis of the Friedmann equation plus ordinary matter and radiation, you get exactly the right answer. If you try to modify the Friedmann equation—for example, try to imagine modifying the Friedmann equation so that the universe is always accelerating—you would be dramatically in disagreement with the predictions of nucleosynthesis and from the microwave background.

All of the data that we have today are telling us that the acceleration of the universe is a recent phenomenon, cosmologically speaking. When the universe was half its current size was the time when the universe was going from decelerating, due to the matter and radiation in it, to accelerating due to the dark energy. If you want to modify the Friedmann equation, you modify it in such a way that the modification turns on relatively late in the universe's history.

Here is the problem with that idea. It can be done, but there are issues that you need to face up to. Here is Einstein's equation again telling us that the curvature of spacetime on the left is related to the stuff in the universe on the right. You want to modify this equation. You want to modify this equation such that, when the stuff goes away, there is still spacetime curvature. Even if there is no energy density, even if the universe expands and dilutes away all the matter and radiation, spacetime curvature will not go to zero. That is your goal.

The problem is that this equation of Einstein's is actually quite remarkable. If you try to mess with it just a little bit, you break it, basically. What we mean by this is if you try to add terms to Einstein's equation that are small, they give rise to some discreet new effects. In particular, they turn on new fields. We mentioned very, very briefly that gravity, the force due to gravity in a quantum mechanical context, can be thought of as being due to the exchange of gravitons, spin to mass-less bosons. But, if you change Einstein's equations in any relevant way, there are more particles than just gravitons. There are new kinds of particles that qualify as part of the gravitational force, and those particles have an effect. What you're trying to do is to mess with Einstein's equation just a little bit, just so that it is important in cosmologically late times when the universe wants to begin to accelerate, but is completely irrelevant here in the solar system. What you find is that that's much harder to do than you thought. At least, the simple things you do have a very interesting effect. We can test general relativity in the solar system and we can ask whether or not the fields that are there, the fields that are affecting the space time metric, are the ones that we're observing.

This is a picture of the Cassini Probe, a satellite launched by NASA, which nominally had as its job to take pictures of Saturn and Jupiter and the other outer planets. But, along the way, it did us a favor. Cassini beamed signals from itself to us here on earth, as it was passing through the path of the sun. We have the sun and behind the sun was the spacecraft. It passed signals to us and one of the predictions of general relativity is that time is warped along with space. As a result of this, the time it takes that signal to get to us is delayed by the gravitational field of the sun. Because we know exactly where the satellite was, you need to plan out its trajectory to very high precision in order to get it to Saturn correctly, we can do a very good job of measuring that time delay, and it turns out it was precisely what general relativity predicted.

What this means is that there aren't any new gravitational fields in the solar system of the kind you would predict if you tried to mess with Einstein's equation. For whatever reason, Einstein's equation is very, very true in the solar system. But, the rule that we have is that if you want to change something on large scales, it's very hard not to change them on small scales. No one has been able to come up with a nice theory where we are able to change Einstein's equation to make the universe accelerate at very, very late times without modifying the dynamics right here in the solar system. You might expect it was possible, but we haven't been able to do it yet.

On the other hand, to take these lemons and to make lemonade, by thinking about these theories, we've been inspired to a new way to test general relativity. What we have in the case of dark energy is an accelerating universe. From the dark energy hypothesis and the amount of acceleration we observe, we can make predictions for how structure should evolve, as the universe gets bigger. We have an early universe, which is very smooth, small, small perturbations and energy density from place to place, which you can see in the fluctuations of the cosmic microwave background. Under the force of gravity, those fluctuations grow into large-scale structure today. You can see in this map, from the Sloan Digital Sky Survey, the distribution of galaxies across the sky. Those come from the tiny ripples that we see in the microwave background. The equation that relates the early, tiny ripples to what we see today is, of course, Einstein's equation.

The point is if you mess with Einstein's equation to make the universe accelerate, you will also be changing the equations that make structure grow. If you try to calculate the effects of dark energy on both the acceleration of the universe and on the growth of structure, and then you go out and observe those two things separately and find that they are incompatible, it is possible to explain that incompatibility by mentioning that gravity has been modified or turning around the same logic. If you find that the exact same dark energy theory explains both the acceleration of the universe and the evolution of structure, then Einstein was right.

Right now we're trying to do this. This is one of the things that cosmologists are embarked on, as we speak, to try to measure not only the expansion of the universe using supernovae standard candles, but also the evolution of structure in the universe. If you get a consistent picture by imagining that there is Einstein's equation correctly and dark energy and dark matter, we'll know that Einstein, who was very smart after all, will probably be having the last laugh.

Lecture Twenty

Inflation

Scope:

Long before we had observational evidence that the universe is accelerating today, we had the theoretical notion that it underwent a period of rapid acceleration at very early times, a scenario known as *inflation*. The physics of inflation is very similar to that of dark energy, including the invocation of dynamical fields with nearly constant energy density. Inflation hasn't yet been demonstrated experimentally, but evidence in its favor is accumulating. Is there any possible connection between acceleration in the very early universe and the acceleration we detect today?

Outline

I. Despite the successes of the Big Bang model, it raises some nagging problems. These are not direct problems with fitting data but problems of naturalness—why is the universe arranged in some particular way that seems rather unlikely to us?

    A. The horizon problem is the most puzzling feature of the conventional Big Bang scenario.

        1. When we observe widely separated parts of the CMB sky, they are at approximately the same temperature; the universe, as we've stressed, is very smooth on large scales.

        2. But we're observing those two points as they were far in the past, because light travels from there to here at a finite speed. And at those times, the universe was much younger—about 380,000 years old.

        3. This is problematic, because the physical distance separating those two points at that time was much larger than 380,000 light-years!

        4. Thus, we're observing two regions that were never in *causal contact*—no signal could possibly have propagated from one region to the other.

5. If they were *never* in contact with each other, the question is, then: how did they possibly "know" that they were supposed to have the same temperature?

B. The flatness problem is a bit more subtle, and experts argue over its importance.

    1. Consider a universe with only matter and radiation, no dark energy. If there is any nonzero spatial curvature in such a universe, its effects will tend to increase in importance as the universe expands.

    2. Thus, we might expect, late in the universe's history, for curvature to be important. But the real world is spatially flat as far as we can tell. Why is that?

II. The inflationary universe scenario uses accelerated expansion in the early universe to resolve the puzzles of conventional cosmology.

A. In 1981, Alan Guth realized that both the horizon and flatness problems could be simultaneously solved by a simple mechanism: a brief period of accelerated expansion in the very early universe, which he dubbed *inflation*.

    1. If the early universe was not dominated by matter and radiation but by some form of (what we would now call) ultra-high-scale dark energy, it would undergo acceleration at a fantastic rate.

    2. This would tend to blow up a tiny patch of space to a fantastically large size, smoothing and flattening it out all the while.

    3. In this way, both the horizon and flatness problems would be solved at once.

B. Inflation quickly became a popular idea, although it was faced with a significant empirical challenge: At the time, observations did not indicate that space was flat! When people measured the amount of matter in the universe, it fell short of the amount required.

C. The discovery of dark energy has provided a great boost to the inflationary universe scenario, for two reasons.

1. First, it demonstrated unambiguously that it was possible for the universe to accelerate, which had not previously been obvious.

2. Second, it accounted for the extra energy density required to make the universe perfectly flat. Thus, a prediction of inflation came true.

D. The mechanism behind inflation—a brief period of accelerated expansion driven by an approximately constant energy density—sounds very similar to the idea of quintessence in the contemporary universe. It's natural to wonder whether they might be connected; right now, the answer is far from clear.

1. The biggest obstacle to a unification of the two ideas is the tremendous separation between them in just about any measure: time, density, energy, temperature—you name it.

2. Furthermore, in between inflation and the present day, we have good evidence from nucleosynthesis and the CMB that the expansion of the universe was dominated by matter and radiation in the conventional way.

3. It seems plausible that inflation and dark energy are completely separate phenomena, but we can't yet be sure.

III. The idea of inflation came with an unexpected bonus—a simple mechanism for generating the primordial density perturbations that grow into large-scale structure.

A. Inflation does its best to smooth out the universe. However, as we've mentioned, Heisenberg's uncertainty principle prevents any state from being perfectly well defined, including the whole universe. Therefore, there is a certain minimum uncertainty in the geometry of space as it expands during inflation; this uncertainty translates directly into a set of perturbations in the energy density.

B. Because inflation happens at a nearly constant rate, it predicts that the perturbations in density should be nearly the same at every length scale.

1. That prediction can be tested precisely against the data; so far, the universe seems to be in good accord with what inflation predicts.

2. It is truly a mind-bending suggestion—that the huge and majestic galaxies we see throughout the universe trace their origins back to quantum fluctuations at literally subatomic scales.

Recommended Reading:

Greene, *The Fabric of the Cosmos*, chapter 10.

Guth, *The Inflationary Universe*, chapters 10–12.

Overbye, *Lonely Hearts of the Cosmos*, chapters 13–14.

Vilenkin, *Many Worlds in One*, chapter 5.

Questions to Consider:

1. The horizon and flatness problems deal with the initial conditions for our whole observable universe. We are, of course, stuck with the universe that we have. Should the nature of the initial conditions count as a scientific question?

2. The ideas of inflation and quintessence are very similar in spirit. Do you think they might be connected? What arguments could you put forward in favor of or against such a possibility?

3. Can we ever know for sure whether inflation happened in the early universe?

Lecture Twenty—Transcript
Inflation

We've been doing a great job of describing what our universe looks like today. We have a model that fits a wide variety of data, but in the last few lectures, we tried to dig in a little bit into that model to the dark energy part of it—the part that is an energy density, that doesn't change from place to place, that is smooth throughout the universe, is persistent as the universe expands, and is more or less constant as a function of time.

What we found is that it is possible to come up with models that explain the dark energy. It could be a strictly positive, constant vacuum energy, one that doesn't change at all. It could be something dynamical that slowly changes or it could even be a modification of general relativity itself. In each one of these cases, it is possible to come up with a version, some sort of model that actually fits the data. But, in none of these cases do we have something that is actually compelling. In none of these cases do we have a specific version, an implementation of this idea, that is not only able to fit the data, but kind of make sense to us—is natural in some way, sort of fits together with other things we know.

Therefore, in this lecture and the ones following, the next two, we're going to step back a little bit and try to think not specifically so much about the dark matter and dark energy, but about the fundamental laws of physics and cosmology—where the universe came from and how it works at a very, very deep level. But, our motivation for doing this is because we want to go back to the dark energy and, to a lesser extent, the dark matter, and understand why these things have the properties they do. Maybe it's not so easy as plugging in something that seems to be something that fits data. Maybe we need to think more deeply about what we consider to be a natural explanation versus an unnatural explanation. In particular, in this lecture, we're going back to near the beginning of the universe. We're going to talk about inflationary cosmology—the idea that the early universe underwent a period of extremely rapid accelerated expansion. Of course, we've been telling you for the last several lectures that the current universe is beginning to undergo a period of accelerated expansion, so the idea that the early universe had a different period of accelerated expansion doesn't sound so crazy. But, *inflation* came on the scene as a physical theory in 1980 and

1981, and back then the idea that the universe could expand was much less accepted. So, it in fact predates the idea of dark energy, the idea that there was a phase of dark energy-like domination in the very early universe, and we call this phase the Inflationary Universe.

Why do we even think about something like this? For our present purposes, inflation is connected to the concepts of dark matter and dark energy in some very specific ways. First, inflation makes predictions and among those predictions is that the universe should be spatially flat. Remember, we talked about the flatness problem, the idea that it would make more sense for the universe to be flat than anything else because it's pretty close to flat already, but inflation provides a dynamical mechanism to make the universe flat. It makes the universe very, very flat, so that's a very strong prediction. Secondly, inflation predicts a certain kind of perturbations to the universe, not only is the universe very flat on all scales, but there are also tiny deviations in the density from place to place. These days, we use those predictions when we're matching what we see in the microwave background, the fluctuations at early times, to the observations we have today of large-scale structure in the distribution of galaxies.

Inflation, of course, is also similar to dark energy, physically. We have some field that is making the universe accelerate, some energy density that doesn't go away very quickly. That's what you need for dark energy. That's also, separately, what you need for inflation, so by thinking about one, you might better understand the other.

Finally, as a spin-off of inflation, we have the concept of the multiverse. We've always known that the universe that we observe right now looks more or less the same within what we observe. Outside what we can observe, we don't know what to say. It could be more of the same, forever and ever and ever, or it could be very different. We just have no way of saying anything.

Inflation at least lets us ask the question scientifically and proposes that maybe the universe is very different outside what we see. That possibility, the possibility of a multiverse where conditions change from place to place in a dramatic way, will turn out to bear on the question of dark energy—the question of why should the vacuum energy be as small as it is, compared to the natural value that it should have. Our expectation for a natural value might be different in a multiverse than it is in a single lonely universe.

So let's think about what people were thinking about back in 1980 when they were beginning to invent inflation and to think about it seriously. The role of inflation is as a model of cosmological initial conditions. The role of initial conditions in cosmology is very different from that of all other physical sciences. If you think about physics as practiced in a laboratory or chemistry or something like that, what you do is you set up an experiment; you set up the initial conditions. You say, I want to have a ball rolling down an inclined plane and I will put the ball at the top. But, cosmology is different than that. In cosmology, you don't get to do the experiment more than once. The experiment is being done as we live right now; the experiment is the whole universe. So, a theory of cosmology, unlike a theory of balls rolling down planes, or a theory of chemistry, contains not only dynamical laws telling you how things evolve, but also a specification of the initial conditions. Why was the early universe in such and such a configuration, that let it lead to the universe which we see today? That is a respectable cosmological question that just doesn't arise if you're doing particle physics or chemistry. So, we have some ideas about what kinds of initial conditions might seem natural or robust or sensible to us. We have other things that seem finely tuned, that seem like for some reason some particular quantity is very, very small when we might have expected it to be big.

Inflation addresses these kinds of questions directly. The entire point of inflation is to make the universe which we actually see at early times appear very natural. With the dynamical mechanism from which you can start in various different types of initial conditions, what inflation does is get you to a point where it looks like our Big Bang universe. The universe looks smooth on very large scales with tiny fluctuations in density, very close to spatially flat. So, let's see how that works.

There are basically two geometrical problems in conventional cosmology that inflation tries to solve. One is called the horizon problem; this is actually the much more serious problem from a cosmology point of view. Think about the cosmic microwave background; think about this image that we can take from satellites. This is from the WMAP satellite, a snapshot of what the universe looked like only 400,000 years after the Big Bang. Now, we know that light travels at a finite speed, one light-year per year, so when we look back in time, we don't see things arbitrarily far away, we see back to the Big Bang. We cannot see any further back than that. Perhaps we can see to something right close to the Big Bang. We don't really know what the Big Bang itself is, but let's be informal for the next few minutes and talk as if we can see to the Big Bang, but nothing beyond that.

The same thing is true for people who would be alive at the time of the microwave background. There was no one alive back then, but imagine an observer sitting at that time in the universe. They would have a past; they would be able to describe certain points in the universe as ones that they can see from signals coming to them at or slower than the speed of light. Other points would have different sets of things that they could see in the universe. But, the thing is, you can do the calculation in a universe that is full of nothing but matter and radiation, no forms of dark energy, and what you find is that in the microwave background, these separated points, in fact any two points that we observe that are more then one or two degrees apart, share absolutely no points in their pasts. In other words, you take one of those points and you extend it to the past all the way to the Big Bang, you take another point of the microwave background extend it to the past all the way, and you get two parts of the universe at very early times that don't overlap.

There is nothing, in other words, in the very early universe near the Big Bang that has the ability to communicate with these separated points on the microwave background. You would have to travel faster then the speed of light. Nevertheless, despite the fact that as we say these points were never in *causal contact*, there is nothing that can get from one point to another slower than the speed of light, these points on the microwave background are very, very close to the same temperature. What that means is these different points in space began to expand at the same time. Nevertheless, they were *never* in contact with each other, so they have their horizons—they have their regions that they can see into the past, that don't overlap. The question is, how do these different points "know" to be at the same temperature? How did these regions of space know to start expanding at the same time? They were never in communication with each other in any way. That is known as the horizon problem.

The flatness problem we've already talked about. If you look at the Friedmann equation of Cosmology, it has three terms. It has the energy density of the universe, the Hubble Parameter (the expansion rate) and the spatial curvature. Basically, if you know what the universe is made of, if you know the stuff inside, let's say it's just matter and radiation—again, imagine there is no dark energy—then you know how the energy density evolves, as the universe grows. You also know how the curvature evolves, as the universe grows. That's just a geometric fact. Therefore, in the Friedmann equation, the Hubble parameter term evolves to compensate. Basically, p, the energy density, and K, the curvature, do what they do as the universe expands, and then H just adjusts to solve this equation.

The flatness problem is the fact that the curvature term goes away more slowly than the energy density term, if the energy density is made of matter and radiation. So, if both the energy density and the curvature are non-zero in the early universe, in the late universe, the curvature should be much bigger, but it's not. If the curvature were exactly zero, it would stay zero and that would make sense. But, why is it exactly zero? Why isn't it some small number or some big number at early times and, therefore, a very big number at late times? That's the flatness problem.

These problems were known in the 1970s. They were known to Alan Guth, who was the one who invented inflation, and inflation turns out to solve both of these problems simultaneously. What inflation says is that you start in the early universe with a tiny little patch of space, dominated by some ultra-high energy form of dark energy. Because it's ultra-high energy, this dark energy accelerates that little patch of universe at a tremendous rate. It's not matter or radiation; it remains approximately constant density and leads to a tremendously fast expansion rate to this little patch of space.

That means two things. One, this little patch of space that might have had a curvature to begin with, has its curvature inflated away by this incredibly fast expansion rate. Remember, curvature goes away; energy density goes away faster, if the energy density is matter and radiation. But, if the energy density is dark energy, it doesn't go away as fast. So, during inflation, the inflationary energy density doesn't go away, but the curvature does. At the end of inflation, that energy density from the dark energy turns into ordinary matter and radiation, which is now much, much, much bigger than the curvature. That's why the spatial curvature in our current universe is so close to zero. It was all inflated away at early times.

The horizon problem is solved because you can imagine an incredibly tiny patch of space—one that was in causal contact and one that did share points in the past and had time to communicate—and inflation takes that patch and expands it to a tremendous size. In other words, inflation changes the past history of the universe in such a way that different points we observe on the microwave background did used to communicate with each other. They were very, very close to each other right at the Big Bang. They did know what each other were doing. There is no horizon problem in inflation.

So, what you need to make that work, of course, is a temporary form of dark energy—a form of dark energy at very, very high energies, that accelerates the early universe at early times and then goes away. In 1980, this was dramatic because we didn't know about our current form of dark energy; this was all brand new stuff. But, these days, we think, OK, that's something we can make sense of.

It was Alan Guth, who was a post-doc in 1980 when he invented the theory of inflation—a post-doc is something in between when you're a graduate student and when you're a professor—you get a series of jobs in which you're supposed to do nothing but write papers and do research, so universities can decide whether or not they would ever want to hire you to be a professor. These days, you might do one or two post-docs before you realize either I have a professor job now or I should find other work. In those days, in the 1970s, you would even do less—maybe one post-doc; two was unusual. Guth was on his fourth post-doc in the ninth year of being a post-doc. Everyone thought that he was really, really smart, so they kept giving him jobs, but he didn't write that many papers, so they didn't give him a professorship position. Finally, he hit the jackpot; he invented inflation. He was actually not trying to solve the horizon and flatness problems when he was working on inflation.

There was another problem called the monopole problem. There was a set of theories called Grand Unified Theories that tried to outdo the Standard Model of Particle Physics, to try to take the strong force, the weak force, and the electromagnetic force that we know and love, and to unify them in a single description. This is a very compelling idea and it still might be right, but it made a prediction at the time that seemed incredibly incorrect. That prediction was there was these particles call magnetic monopoles, individual magnetic charges that we don't see in nature, but we do see in individual electric charges. According to Grand Unification, the universe should be full of magnetic monopoles, but we don't see any. How do you get rid of them was the question that Guth was trying to answer.

Not only does inflation solve the horizon and flatness problems, it also solves the monopole problem. If you have a high density of monopoles in the early universe, all you do is you inflate them away. Since then, inflation has become a great cure-all for anything that the early universe creates that you don't see in the later universe. As long as those things that were created were created before inflation, inflation can dilute everything away by a tremendous amount before its dark energy turns into ordinary matter and radiation.

Guth realized that he had a solution to the monopole problem; he already had in the back of his mind knowledge about the horizon and flatness problems, and realized all at once that his idea of inflation solved them all. He literally late at night was working and in his notebook wrote, "spectacular realization" and put a box around it, that his idea of inflation could solve not only the monopole problems, but also the horizon and flatness problems. That notebook that he was writing in is now on display in the Adler Planetarium in Chicago. It was a moment in the history of cosmology when we realized that this one idea could solve a whole bunch of problems all at once. People realized this; people caught on very quickly to the idea that inflation was a great help to the various different cosmological conundrums that we had. Of course, among other things, Alan Guth got a faculty job very quickly; he's now a full professor at MIT.

One of the nice things about inflation was that it provided predictions. It was a scientific theory that made predictions that could come true or be false. Its strongest prediction was that the universe should be spatially flat, that the total energy density of the universe should be the critical density. This is an interesting prediction because it was made in 1980, and throughout the '80s and most of the '90s, it didn't look like it was true. People thought that there was enough uncertainty that maybe the universe did have the critical density, but as they measured more and more about the energy of matter, they found out it wasn't enough. They honed in on the energy density and matter being about 30 percent of the critical density.

So, there are two really important things that dark energy does to help the idea of inflation, to boost our confidence that something like inflation is right. First, most obviously, the dark energy provides the extra 70 percent of the density of the universe that we need to make the universe spatially flat. In other words, by 1997, if you believed in inflation, you were worried. There were some people who were actually backsliding and trying to create models of inflation that had universes that did not have the critical density, that were negatively curved spaces instead of flat spaces. Guth himself never actually went that far. You can invent such models, but they are incredibly ugly. The true prediction of inflation is that the universe should be spatially flat.

So, in 1998, when the supernova evidence came in that there was such a thing as dark energy and you could make a spatially flat universe without only relying on matter, both ordinary and dark, it made the case for inflation much stronger. That was a prediction of inflation that came right. Then, in 2000, when boomerang and other microwave background experiments said yes indeed the universe is spatially flat, inflation was, of course, right on.

The other idea that was helpful to inflation from dark energy is the very demonstration that the universe is allowed to accelerate. Remember, we have something called the cosmological constant problem. Why isn't the energy density of the vacuum much bigger than we apparently observe it to be? We don't know the answer to that problem but, before 1998, it was always possible that the answer was that the vacuum or other forms of dark energy do not gravitate. There was something in the laws of physics that says that the expansion of the universe just doesn't respond to things with negative pressure. No one had a good model along those lines, but it was an allowed way to think. If that had been true, it would be difficult to understand how the universe could possibly accelerate. So, the fact that we are now observing the universe to be accelerating right now means that it is allowed to accelerate; therefore, it could have been accelerating at early times when inflation was necessary. In other words, inflation is on much better physical grounds now than it was before.

How do you make it work? How do you invent a model of inflation? For dynamical dark energy, we invented a field call quintessence that slowly changes its energy density as the universe expands. Exactly the same thing is true for inflation. You invent a new field; you call it the inflaton because you have no idea what it is. It's the field that makes inflation happen. It's a field that has a huge energy density, at very early times, and becomes the dominant form of energy in some patch of space, which then accelerates, inflates at a tremendous rate. This happens in the energy density and that inflaton field gradually diminishes, but very, very gradually, so that the expansion rate is continually accelerating. Then at some point there is a transition. There is a phase transition where the energy density in that dark energy transforms into ordinary matter and radiation. We call this reheating. In other words, it's a nearly constant energy density for a while, and then it snaps, turns into matter and radiation that we know and love. We see that as the Big Bang. That's the basic idea of inflation.

So, because the physics underlying inflation sounds to our ears very similar to the physics underlying quintessence or dynamical dark energy, some people have asked the question, is it in fact exactly the same thing? In other words, is there one field that was the inflaton, at very early times, provided the dark energy back then, and also is now the quintessence field providing the dark energy right now? Papers are written with titles like Quintessential Inflation. The opportunity for a pun in this field is never passed by.

The answer is, it could happen; it could be that the same field is responsible for inflation and for the dark energy today, but probably not. For one thing, the energy scales are tremendously different than each other. The energy density of the universe, near the Big Bang, when inflation was going on, was many orders of magnitude higher than it is today. It is possible that the energy density was dominated by the same field then and now, but what you have to do is make that disappear in between or at least be dramatically subdominant. At least from the time of Big Bang nucleosynthesis to the time of today, just before today, the universe was certainly dominated by ordinary matter and radiation. By ordinary, I mean by matter-like particles, including dark matter.

We know from the data, from Big Bang nucleosynthesis, and from the cosmic microwave background, that the universe wasn't dominated by dark energy all along. The dark energy has kicked in recently. Inflation says the dark energy was dominating way back then, so it's actually easier to make those two periods of domination be due to two completely different fields than to the same field that is important back then, disappears, and then comes back. It's hard to make one field be so different that it dominates at very high densities and at very low densities. But, it's still the kind of idea that people are working on. It might end up being right. We'll have to go see.

The other thing that inflation gives us is a bonus. Not only does it explain away the horizon problem, the flatness problem, and the monopole problem, but it gives us a dynamical origin for the density fluctuations we observed in the universe. If you think about it, the universe on very large scales is in a very strange state. It's very, very smooth on very large scales, but not perfectly smooth. The deviations from smoothness that we can observe, 1 part in 10^5 are certainly observable; they are not absolutely absent. So, if you think about it, why would it be that the early universe would undergo some process that made things very, very smooth, but not perfectly smooth? Why isn't it either very, very lumpy or even smoother? The answer in the context of inflation comes down to quantum mechanics. Inflation tries to expand the universe and smooth it out. As the universe expands, is it is accelerated by some form of dark energy. That acceleration smoothes out bumps and ripples. But, quantum mechanics and the uncertainty principle say that you can't smooth out everything perfectly. You're trying your best to make the universe smooth, but the field that is doing it, the inflaton field that is driving the energy density, has quantum mechanical fluctuations, a little bit of jitteriness that you can never get rid of.

It is those quantum mechanical fluctuations that turn into perturbations in the density of matter and radiation in dark matter, and those show up in the microwave background as temperature fluctuations and they grow into galaxies today. In fact, there is a prediction on top of that—namely, that the amplitude of the fluctuations should be more or less the same at every distance scale. That because inflation, as it's happening, is happening at more or less the same rate at every distance scale and it is imprinting fluctuations as it goes along. This is of course exactly what we do observe when we look at the microwave background. When we look at large scale structure in the universe, we see perturbations that seem to be about the same primordial amplitude, whether they are one parsec across or on gigaparsec across, so inflation at least is coming close to being correct with that prediction.

The other bonus that we get is a little bit less tangible. The tangible bonus that we get from inflation is the fact that there are density perturbations that are predicted and, in fact, it has become a cottage industry amongst cosmologists to think about how inflation leads to those perturbations, where they can come from, and how you can test them. There is a slightly more speculative outcome of inflation, which will turn out to be useful two lectures from now, which is the multiverse. We alluded to this in the very beginning. We have this picture of inflation in which there is a tiny patch of space at early times, it was dominated by the dark energy in an inflaton field, and accelerated at a tremendous rate, grew up to be a universe size thing, what we live in today. But, back then, what about the other patches of space? What about the other parts of the universe, which were not in the little patch that initially inflated to become our universe? It's easy to imagine that there were all sorts of fluctuations, that the universe was very different from place to place, way back then, before inflation ever happened.

So, inflation grabs this little piece of the universe, expands it into what we see today, but other regions could easily be grabbed by different kinds of inflation, different inflaton fields, or just the same inflaton field, but evolving in a different way. The act of inflation might very well be to take different parts of the universe and to blow them up into universe size pieces, but end up in very different conditions. So, we need a theory of how the universe could be in different conditions, but inflation allows us to talk about, in a scientific way, the possibility that outside our observable universe, conditions are very, very different. That will change how we think about what constitutes a natural value for things like the vacuum energy, other constants of nature. We will have a whole multiverse and the possible existence of a multiverse will affect how we think about problems involving both dark matter and dark energy. We'll get to that in a little bit.

The important thing then, once we're stuck with the idea of inflation or once we are granted the idea of inflation, which is a great idea, how do we know, number one, whether it's true, number two, how do we make it work? These are exactly what cosmologists today are thinking about in very serious ways.

First, we want to test the idea of inflation. We said that inflation makes a prediction. It takes a little patch of universe, expands it enough to be spatially flat, so indeed we will expect a spatially flat universe. It also makes specific kinds of predictions about density fluctuations. In fact, it predicts that the amplitude of density fluctuations should be the same at every distance scale. Both of those seem to be more or less true in the universe we observe. However, both of those were conjectured to be true even before inflation was invented. The problem with these predictions, the universe should be spatially flat and the perturbations should be the same amplitude on different scales, is that they are very vanilla. They are not very flavorful. You could imagine that these are the simplest possibilities for what is true, even if you don't have a mechanism for inflation that makes them true. So, people imagine that the universe was spatially flat, people imagine that there were different density fluctuations with the same size on different physical length scales, before anyone ever invented inflation. So, even though they are predictions, they are not unique predictions. One could certainly imagine that they are true, without inflation being true.

What we want to test inflation is something more unique, something that inflation says is true that other theories don't necessarily say is true. There is one known example of such a thing. Inflation, when it's expanding the universe, has the inflaton field itself and its quantum fluctuations. Those quantum fluctuations get imprinted into density fluctuations, and grow under the force of gravity into galaxies and large-scale structure.

But, there is another field lying around at that time when inflation was going on—namely, the gravitational field. The gravitational field will also undergo small quantum fluctuations during inflation. These quantum fluctuations will get expanded to be the size of the universe. So, what is the observable form of a small fluctuation in the gravitational field? The answer is a gravitational wave, gravitational radiation, or individual gravitons. Gravitational radiation is something we are looking for in laboratory experiments here on earth and haven't yet found. But, what inflation says is that there should be a background of gravitational waves filling the universe. It is possible or at least conceivable that these gravitational waves could be detected by certain special kinds of observations of the cosmic microwave background.

Remember that when we look at the microwave background and measure its temperature, it changes from place to place. But, since we're measuring photons, we can also measure the polarization of the cosmic microwave background. If the inflationary prediction of gravitational waves is true, there will be a very specific kind of imprint on the polarization of the cosmic microwave background. We have so far detected that there is polarization of the cosmic microwave background, but our current measurements aren't good enough to test the prediction of inflation. This is one of the things that we're shooting for in future generations of experiments. To improve exactly those measurements, we might be able to verify that either inflation or something very much like it was true in the early universe.

That will still leave us with questions. Even if we know that inflation happened, we're still left with the question of how it happened. What was the universe really like before inflation began? How did inflation start? Why did that little patch become dominated by dark energy? For that matter, what is the inflaton? What is the field that was responsible for the energy density that led to the acceleration we witness in inflation? All of these are good open questions. The good news is that we have ideas for them. The bad news is that they are a little bit past our reach in terms of experimental capabilities. So, what we need to do is to think more cleverly about what could have been going on at the time of inflation and, in particular, think about how to apply those ideas to things we can observe and test in the universe, so that we can turn inflation from a promising speculation into an established part of our understanding of the early universe.

Lecture Twenty-One
Strings and Extra Dimensions

Scope:

String theory is an ambitious attempt to unify gravitation with the other forces of nature. It is a candidate for a *theory of everything*, in which *everything* includes dark matter and dark energy. Unfortunately, connecting string theory with the observed world has proven difficult; for one thing, strings naturally move in more than three spatial dimensions, and the extra dimensions must somehow be hidden from our view. These extra dimensions can play a crucial behind-the-scenes role in governing how physics works in our observable four-dimensional spacetime.

Outline

I. Quantum mechanics and gravity will ultimately need to be reconciled.

   A. As a matter of practice, the Standard Model combined with general relativity does a wonderful job of accounting for essentially all our experimental knowledge about how the universe works, at least here on Earth. But as a matter of principle, these two frameworks are fundamentally incompatible.

   1. Particle physics operates within the framework of quantum mechanics, while general relativity has stubbornly resisted incorporation into that picture.

   2. Reconciling general relativity and quantum mechanics is perhaps the single most important problem in fundamental physics today.

   B. There are conceptual obstacles to coming up with a working theory of quantum gravity, but there are also direct technical problems. Consider the scattering of two particles interacting via gravity.

   1. If we draw the simplest Feynman diagram describing such a process, it gives a sensible answer, which should be a good approximation to the exact answer. To improve the approximation, we start including somewhat more complicated diagrams.

2. When we calculate the contributions corresponding to these diagrams, however, we find that they are infinitely big! This result is not at all a small improvement to a sensible answer.

3. Sometimes, in the course of doing quantum calculations, we find infinite answers that aren't that worrisome—they can be "renormalized" into finite answers. For quantum gravity, that doesn't seem to be the case; the answers simply don't make sense.

II. String theory is the leading candidate for a consistent theory of quantum gravity.

A. String theory has a daunting reputation, but at heart, it's an extremely simple idea.

1. In ordinary particle physics, we imagine that every particle is a perfect mathematical point. If you were to zoom in ever closer with a microscope, it would remain perfectly pointlike.

2. String theory simply says that this process actually does not continue forever.

3. If you look closely enough (much more closely than we currently have the ability to do), particles resolve themselves into tiny loops of string. These strings are approximately the Planck length (roughly 10^{-33} centimeters); that's much smaller than the size of a proton (about 10^{-13} centimeters).

B. You can't ask what the strings are made of—they're made of string stuff; that's as deep as it goes. But from that one simple idea, many consequences flow.

C. One consequence is gravity. Whereas ordinary field theories seem to be incompatible with gravity, string theory necessarily predicts it!

1. One of the vibrational modes of a string, like it or not, is a spin 2 massless boson with all the couplings of a graviton.

2. Furthermore, the extended nature of the string works to smooth out the infinities that are so troublesome in naive approaches to quantum gravity. Not only is string theory a quantum theory of gravity, but it's a well-behaved (finite) one.

D. The bad news is, as attractive as string theory is in principle, it's extremely hard to connect it to practice, for reasons we'll see in a moment.

III. Extra dimensions are a prediction of string theory.

A. How do we know that three dimensions of space (and one of time, of course) are all there are?

    1. One way is to tie sticks together; you can tie three sticks together so that each one is perpendicular to the other two, but you can't do the same trick with four.

    2. However, in the dynamical spacetime of general relativity, we can imagine that there are actually extra dimensions of space but that they are curled up into tiny spaces too small for us to notice.

    3. Indeed, this idea was pursued by Theodor Kaluza and Oskar Klein soon after general relativity was first invented.

B. Extra dimensions remained an amusing possibility for a long time. But the emergence of string theory brought them back with a vengeance, for a simple reason: Although extra dimensions are optional in GR, in string theory, they are mandatory.

C. Strings require the existence of extra dimensions; to make the idea compatible with our observed world, they must somehow be hidden.

    1. The simplest way is to follow Kaluza and Klein and simply curl them up into an invisibly tiny space.

    2. But string theory is not just a theory of strings; it also predicts larger objects extended in one or more dimensions—objects known as *branes* (derived from the word *membrane*). Imagine a brane with three spatial dimensions, embedded in a larger space of more dimensions.

a. An interesting feature of branes in string theory is that certain types of strings (and, thus, certain types of particles, from our macroscopic point of view) can be confined to the brane.

b. We can imagine, in fact, that all the particles of the Standard Model that we know and love are actually confined to a three-dimensional brane in some larger universe.

c. In such a scenario, we don't notice the extra dimensions simply because we can't get there—we're made of such particles ourselves, after all.

D. The upshot of this story is that there are many more ways to hide the extra dimensions than we previously suspected. We don't know exactly how many, but such numbers as 10^{500} are routinely suggested.

1. Every one of these ways gives rise to a macroscopic world with a different set of particles and forces.

2. Just as water can come in different phases—solid, liquid, and gas—so can spacetime itself, according to string theory. Except that, instead of three phases, there are 10^{500}.

E. All these different phases, unfortunately, represent a significant obstacle in our attempts to tie string theory to the real world. We should live in one of these phases, but which one? Gravity is a very weak force, which makes it difficult to devise direct experimental probes of any theory of quantum gravity; we hope that we will eventually be clever enough to think of a way to test the theory experimentally.

Recommended Reading:

Greene, *The Fabric of the Cosmos*, chapters 12–14.

Randall, *Warped Passages*, chapters 14–16.

Questions to Consider:

1. It's hard to think of ways to test string theory experimentally. On the other hand, both gravity and quantum mechanics are part of physics and should eventually be reconciled. How far can physics go on the basis of thought experiments alone?

2. Imagine that we did find experimental evidence in favor of string theory and derived a unique model connecting it to the real world. Would that be the end of physics?

3. Critics of string theory have suggested that it has occupied the efforts of too many bright young physicists, who find it difficult to strike out with their own radical new ideas. How can we balance work on a promising but speculative approach with attempts to try something completely different?

Lecture Twenty-One—Transcript
Strings and Extra Dimensions

Dark matter and dark energy, taken together, are about 95 percent of the stuff in the universe. However, they represent much less than 95 percent of the lectures you can buy from The Teaching Company. Part of our goal in this set of lectures is to sort of redress that balance to really concentrate on that 95 percent, on the dark matter and the dark energy.

When we talked in the last lecture about inflationary cosmology, it's not because inflation gave us directly a theory of dark matter and dark energy. Rather, it relates to the background set of assumptions that we deal with when we start talking about things like dark matter and dark energy. When we are cosmologists working and trying to understand the current universe, the idea that the universe probably underwent a period of inflation at very early times colors how we think about everything else. When we discover new things about the universe, it colors how we think about inflation. For the obvious example, when we discovered dark energy, it helped show the prediction of inflation that the universe was spatially flat was something that was going to be correct. Likewise, that physics of inflation was a scalar field, a field that is slowly rolling and changing its energy density only very gradually, has showed up once again in ideas for dark energy.

Today, we're going to be talking about a different background theory, the string theory. String theory is what a lot of people have in their minds in the background when they are thinking about the fundamental laws of physics—both particle physics, as we know it in the Standard Model and perhaps beyond, and also gravitational physics, as understood in general relativity. Like inflation, string theory is a speculation. It's not something that we absolutely know is true, but it's something that has caught on as a very popular speculation. There is a tremendous amount of intellectual effort going on right now trying to relate the ideas of string theory to the things that we observe here in nature.

The idea behind string theory is really extremely simple. It's just a statement that you replace elementary particles with elementary strings. In other words, in a conventional way of looking at things, if you take an electron, or a quark, or a photon, and you zoom in on it with the most powerful microscope you can imagine, they still remain pointlike. They don't have any extent in any direction. They are just fundamental geometric points. So, string theory says that's not right. Instead, these particles that we think are particles are actually little loops of string. That is to say, they have one dimension of extent that takes the shape of a circle, and that circle is vibrating just a little bit. We don't know what the strings are made of. It's not even a sensible question to ask. They are made of the stuff that strings are made of. But, the idea is the different ways you can vibrate a string correspond to all the different particles that we see.

This theory has caught on for many reasons, primarily because it's a promising theory of quantum gravity, as we'll discuss. Also, it's prospectively a *theory of everything*, and *everything* includes dark matter and dark energy. What we're going to be talking about is the possible ways that the ideas behind string theory can influence our conception of dark matter and dark energy.

As far as dark matter is concerned, there is sadly more than one possibility for the ways that string theory can give us a candidate dark matter particle. It's consistent with our ideas of supersymmetry and WIMPs, also with our ideas of axions and perhaps even neutrinos. For dark energy, it's a trickier situation. You can get quintessence out of string theory, but it's not very natural. The very intriguing suggestion is that string theory can offer an explanation for the size of the vacuum energy; that's the good news. The bad news is that the explanation involves 10^{500} different regions of spacetime, which we don't see. We're going to evaluate that suggestion very critically in the next lecture. In this lecture, we're going to try to set up what string theory is, why we're motivated about it, what the current state of the art is, and try to connect it to other things about the world.

The reason why we care about string theory is mostly because it's a prospective theory of quantized gravity. One of the things that we've had all throughout these lectures is that, in physics, all the theories we have, and all the data that we have, have to be tied together. We don't have a separate theory for every phenomenon. You have the smallest possible set of theories that can explain the greatest possible set of phenomena. In particular, you know that if your theories are really going to be right, simultaneously, they need to be consistent with each other. If you have two ideas that work really well in their respective domains of validity, but are fundamentally incompatible with each other, then you know that something has to give. You know that either one of them is wrong and needs to be replaced, or you need to find a third theory that takes both of them into account and reduces to two different examples in two different limits. This is a way that scientists can make progress even in the absence of direct, explicit, and specific experimental data. Sometimes you know your theory is wrong because you do an experiment and it contradicts the theory. That's an easy way to know that there is something that needs to be fixed about how you're thinking about things. But, sometimes you know your theory is wrong even though it's consistent with all the data you have, and that is the situation we find ourselves in right now.

It's not the first time. In the past, this has worked. For example, in 1900, we had the theory of electromagnetism as put together by Maxwell with Maxwell's equations. We also had the theory of classical mechanics that was put together by Isaac Newton. But, these two theories were not quite compatible. They had different sets of symmetries. They gave different answers to what would happen if you observed the same set of phenomena in two different reference frames. So, just by thinking how to reconcile these two points of view, Einstein was able to come up with his special theory of relativity. Almost immediately after inventing the special theory of relativity, he realized that this new theory was incompatible with Newton's theory of gravity. He would once again have to reconcile these two things. There was some experimental guidance. He knew about the orbit of Mercury being an anomaly. Nevertheless, his primary goal was to find a single framework, which could reduce, in the right circumstances, both to Newtonian gravity and to special relativity. He ended up coming up with general relativity.

Now we have a giant looming incompatibility in the fundamental laws of physics. As far as we know, every experiment we've ever done—here on earth, at least—can be explained by one of two theories: The general theory of relativity, Einstein's theory of curved spacetime as gravity, or the Standard Model of Particle Physics, a set of fields and their interactions governed overall by the rules of quantum mechanics and quantum field theory. The problem is general relativity is not a quantum field theory. It's a classical field theory. Even though, in some sense, general relativity replaced ideas

©2007 The Teaching Company.

of spacetime that we got from Isaac Newton, in philosophy, general relativity is very Newtonian. It says there is a state of the universe, there are equations that govern the evolution of that state, and we can observe that state in any way we like, in principle, to arbitrary precision.

The rules of quantum mechanics say something very different. They say that we can't observe everything about a system. There is a wave function. A wave function tells us what the system is really like, but when we observe it, we don't observe the wave function directly. There are certain observable features of it and when we look at something, the observable thing we get depends on the amplitude of the wave function. What the wave function is really telling us is the probability for getting certain results when we actually observe something.

We want to reconcile these two theories. It can't be the case that general relativity is correct and that quantum mechanics is correct. If general relativity is correct, then the source of gravity is energy and momentum. But, quantum mechanics is telling us that energy and momentum are just things we observe. They are not the fundamental things that are. In other words, we could imagine, at least hypothetically, a quantum mechanical situation in which you have a gravitating body and the wave function of the gravitating body says there is a 50 percent chance it's over here and there is a 50 percent chance that it's over there. Try to ask the question, according to classical general relativity, in what direction does the gravitational field point? Does it point towards there or does it point towards there? Hopefully, the obvious answer should be that there is a 50-percent amplitude that it's there and a 50 percent amplitude that it points there. But, what that means is that spacetime itself needs to be quantized. We have a wave function of the whole universe, a wave function of the dynamical properties, and the curvature of spacetime itself. That is what we seek in a quantum theory of gravity, and we don't have it right now.

It turns out, of course, that we have a set of cookbook procedures for taking a classical theory and quantizing it. The real world we think is quantum mechanical. It's not really classical and quantum mechanics is something we add on top. It's really quantum mechanical and classical mechanics, à la Isaac Newton, is something that is a limit, an approximation to the true quantum mechanical realities that work very well when objects are really, really big. When you're working with individual atoms or electrons, quantum mechanics is absolutely necessary, but when you're working with big objects like the earth, or the sun, or even you and me, classical mechanics is perfectly fine.

There is no reason, ahead of time, that the way to find the correct quantum mechanical theory of reality starts by taking the classical theory and then quantizing it. Nevertheless, it turns out to work awfully well in many different circumstances. The most successful theory that we have within the realm of quantum field theory is *quantum electrodynamics* (QED). This is what won Nobel Prizes for people like Julian Schwinger and Richard Feynman. It is the quantum theory of electrons interacting with photons, and it works extremely well.

But, we have a classical theory of electromagnetism that also works extremely well. The way to get quantum electrodynamics is to quantize classical electrodynamics. When you try to take that cookbook and apply it to general relativity, it works at first and then it breaks down. It's not as if we don't understand anything at all about quantum gravity. We can take the first few steps. For example, you can consider gravitational fields that are weak. Spacetime is almost flat everywhere with tiny ripples on top of it. As a matter of fact, in most of the observable universe, the gravitational fields are actually pretty weak, so this is a very good approximation for many circumstances. We can quantize weak gravitational fields very easily. We get a theory of gravitons. We get a theory of particles that are quantized excitations of the gravitational field, in exactly the same way that when we quantize electromagnetism, we get a theory of photons, the quantized excitations of the electromagnetic field. So, the theory of gravitons works pretty well. It makes sense. We can scatter gravitons off each other. But, then, we try to push this theory a little bit harder. We say what about when the fields aren't weak? What about what happens at very, very small distances? There, it breaks down. What we have are two kinds of problems. We have technical problems; we also have conceptual problems.

The conceptual problems are that if you have a background spacetime on which there are fluctuations in the gravitational field, you can quantize those fluctuations and the background spacetime is still more or less classical. But, if you want to truly quantize spacetime itself, it becomes a question of what are we doing? In ordinary quantum mechanics, for example, the wave function depends on time. You tell me what time it is, I solve Schrodinger's equation, and I tell you what the wave function is. But, now, in general relativity, time isn't out there in absolute. It's not a background in which we move. It's part of spacetime. It's part of the thing we're quantizing itself. So, should the wave function depend on time? It's more likely that somehow time emerges out of the correct understanding of quantum gravity. But, even though I can say all of those words, I don't know how it actually works. This is one of the things that we're trying to address, one of the obstacles we have right now into making quantum gravity into a sensible theory.

But, even without those conceptual problems, we have technical problems. That is to say, we try to do what we think we know how to do and we run into nonsense. There is a famous saying that often happens in attempts to quantize field theory, which is you get infinitely big answers. You can take two particles, you can scatter them, and you can ask, according to the rules of quantum field theory, how likely is it that these particles actually hit each other and bump off? If your field theory is well behaved, like quantum electrodynamics is, you get a finite answer. There might be steps along the way in which it looks infinite, but at the end of the day, you get a very well behaved finite thing.

For some theories, that doesn't happen. For some theories, your first approximation to the answer makes perfect sense. Your second approximation is infinity. Things break down and there is no known way to fix them. Usually, in nature, when this happens these infinities are a sign that new physics is kicking in that you don't understand. For example, Enrico Fermi's original theory of the weak interaction had this property. The first approximation worked well; the second approximation was infinite. That was the problem. But, now that we understand W and Z bosons, now that we understand the correct theory of the weak interaction, that problem has gone away. So, this problem exists with gravity. When you scatter two gravitons off each other and look closely at what the answer is supposed to be, it looks infinite. We, therefore, suspect that there is new physics kicking in, the equivalent of W and Z bosons, but for gravity. That is the kind of thing that string theory purports to offer us—the new physics that makes everything suddenly make sense and gives finite answers.

So, string theory is just the idea that everything is made of incredibly tiny loops of string. If you were able to zoom in with a super-powerful microscope on what was going on at very, very small length scales, you would be able to resolve individual things that you thought were particles into vibrating loops of string. The reason you're not able to do that is because the loops of string are extremely, extremely small. They are approximately the size of the Planck length, a length that was invented by German physicist Max Planck to describe the scale in which quantum gravity begins to become important. In numbers, it's about 10^{-33} centimeters across. That sounds like a very small number, and it is. In fact, we can compare it to the size of a proton. A proton made of three quarks held together by gluons is small. It's smaller than an atom by quite a factor, but it's still about 10^{-13} centimeters across. In other words, the Planck scale is 20 orders of magnitude smaller than a proton. It's a much tinier length than any of our current particle accelerators are giving us access to. If you look at these tiny vibrating loops of string, you will see them as particles. They look particle-like to us because we can't resolve the fact that they are little one-dimensional loops of string.

When we talk about what string theory predicts, we will still be talking as if it predicts different kinds of particles. There will still be protons that will be particles, etc. But, they will really be different vibrational modes of the string. That is the idea behind how string theory is able to give rise to all sorts of different kinds of particles. It's not that the stringy stuff is made of different kinds of stuff or different kinds of particles. There is just one essence of string stuff that is the same for everything. The little loops of string can vibrate in different ways and the different modes of vibration correspond to bosons, fermions, what have you.

Interestingly, there are, in fact, many ways to start with those vibrating strings and get a consistent theory. In fact, since the 1980s when string theory first became very popular, it was felt that there were only five possible ways to start with a little vibrating loop of string and get one consistent theory. These days, as we'll talk about briefly, we think that there is only one way to do it. Not that we've said that the other ways are inconsistent, but we've found that what we thought were five different string theories are actually just five different versions of the same underlying theory called M theory.

This is part of the charm of string theory; that you start with an incredibly simple idea. There is just a loop of stuff and you quantize it. How simple could that be? There is really nothing else there. I'm not leaving out any hidden assumptions. You find a unique theory somehow. You find particle physics and gravity, everything you want to find just out of that little loop of string.

In actually connecting it to the world, we try to make predictions. We try to say, how do we relate string theory to particle physics? The first thing, of course, is gravity. The reason why string theory was invented in the first place was as a theory of the strong interaction. In the late 1960s to early 1970s, we didn't yet put together what we now call QCD, *quantum chromodynamics*—the correct theory of quarks with different colors held together by gluons. We were still in the late 60s guessing at what the correct theory of the strong interactions really was.

One person noticed that if you looked at all the different particles that seemed to be relevant to the strong interactions, they arranged themselves with their masses and their spins, as if they were little vibrating loops of string. String theory was born as an attempt to explain the strong interaction. But, it didn't work. One of the reasons why it didn't work was because it kept predicting a massless, spin-two boson that coupled to every form of energy and momentum. In other words, string theory kept predicting a graviton; therefore, it'd be a theory of gravity. Even though you didn't want gravity, it kept appearing. This is kind of a miraculous thing. Ordinarily, when we try to quantize gravity, we run into trouble. We run into infinities, or conceptual problems, or whatever. String theory is an example of a model where you didn't want gravity. You were trying to explain how protons hang together and you kept predicting gravity. It turns out that not only do you predict a quantum theory of gravity, but it's a well-behaved theory.

One way of thinking about why you get these infinite answers in ordinary attempts to quantize gravity is because the Feynman Diagrams become exceptionally complicated. Feynman Diagrams describe the different ways that particles can come together and interact, but to a particle physicist, each one of them gets attached to a number. You can get complicated loops inside the Feynman Diagrams. As those loops become small, the number attached to the diagram can become very large, possibly infinite.

What happens in string theory is that you replace the Feynman Diagrams with just lines connected to each other with the paths of strings through spacetime, so they look like two-dimensional sheets. You have a one-dimensional string moving through time that describes a two-dimensional sheet moving through spacetime, which could describe a single particle splitting into two, for example. The number of possible diagrams you can draw describing such strings is, number one, much smaller than the number of Feynman Diagrams that you can draw, and number two, much simpler. It turns out that what string theory does is to smooth off the sharp corners that would be there in the Feynman Diagrams of ordinary quantum gravity and you begin to get finite answers. Even better, you get more than a theory of quantum gravity; you get possibly a theory of everything. The different ways the strings can vibrate give rise to gravitons, but also to photons, also to gluons, to electrons, and quarks. All the kinds of particles we know about in the Standard Model can be reproduced by vibrating strings. That's the good news. The bad news is that to actually reproduce them specifically turns out to be very, very hard.

It turns out to be very difficult to go from what we think is the unique theory of string theory to the world in which we live. That's not really surprising because, in one sense, gravity is very weak. Quantum gravity is something that is very far away from our experimental reach. We don't have a lot of explicit clues about how gravity and particle physics get together. We're left with only our IQ points. Sometimes they are good enough; sometimes they are not up to the job.

What we're trying to do is to get string theory and match it to the real world, so we ask what predictions does string theory really make? The one prediction that string theory makes more strongly than anything else is that spacetime is 10-dimensional. Strings wants to propagate in a space, which is 9-dimensional, and one dimension of time. The bad thing about that is that it's not the real world. At least, the good thing is that it is a prediction of something, and the question is, is it possible to reconcile 10-dimensional spacetime into the four dimensions that we see? The answer there is actually yes. It's been known for a long time how to go from a higher dimensional space to a lower dimensional space. If it had been the other way around, if string theory only worked in a two-dimensional spacetime,

©2007 The Teaching Company.

we would truly be in trouble. But, it's easy to get rid of extra dimensions of spacetime.

String theory predicts there are six dimensions of space that we don't see. The way to get rid of them is to curl them up. This is an idea that goes all the way back to Kaluza and Klein, soon after Einstein invented general relativity. They said that if there are dimensions of spacetime that are dynamical and curved, maybe there are ones that we don't see because they are curled up into tiny little circles or tiny little spheres. So, string theory takes this idea and it tries to take advantage of it. It says maybe these extra dimensions that we don't see really are there, but they are curled up into little balls the size of a Planck scale. That's the good news. Furthermore, the different ways in which you can compactify those extra dimensions tell us what the particle physics will look like at low energy. In other words, if string theory is right, the reason why, for example, we have three families of fermions in the Standard Model turns out to be a particular topological fact about the space on which we compactify the extra dimensions. We turned a statement about the number of particles we see into a geometrical statement about extra dimensions that we don't see. That's interesting because it offers a new way to solve the kinds of problems we were stuck with—why are all these particles there?

The bad news is that there are too many ways to curl up the extra dimensions. The ways we have to curl them up are not unique. From the mid-1980s when string theory first became popular to the late 1990s, people were crossing their fingers and hoping that somehow there would be one way to compactify that was the best. People were trying to look for that and what they found was that with different models of compactification, they could get all sorts of different possibilities.

For example, supersymmetry is a robust prediction of string theory, as well. That's good news if we're thinking about dark matter because our favorite theory for the dark matter might be the lightest supersymmetric particles. But, then, it turns out that string theory also predicts that there should be axions. It also predicts that neutrinos should have mass. So, so far, string theory is predicting too many things. It's not picking out one candidate for dark matter as the right one.

Things got slightly worse in the 1990s when it was realized that string theory is not just a theory of strings. Remember that one of the virtues of string theory is that it's unique. You start at a particular place and you derive things that are necessary. You can't get around them. The non-uniqueness comes when you go from a large 10-dimensional world and compactify it. There are many, many ways to do that. But, in the original 10-dimensional description, things are very unique. People realized that in that unique description there lived not only one-dimensional loops of string, but higher dimensional objects known as *branes*. The word "brane" comes from the word *membrane*, which means a two-dimensional thing. If you live in three spatial dimensions, the only things you can really easily imagine are zero-dimensional particles, one-dimensional strings, and two-dimensional branes. But, if you have 10 dimensions or nine dimensions of space to play with, then you can imagine three branes, four branes, five branes that are three-dimensional, four-dimensional, five-dimensional, etc. It was discovered in the 1990s that all of these do play a role in string theory. There are different kinds of branes; some of them have miraculous properties. Some of the miraculous properties are that there can be particles, strings, fields, that are confined to the brane. In other words, you could have a brane living in some higher dimensional space, but all the electrons and protons of this particular construction were stuck on the brane. They can't escape out into the extra dimensions. So, if you were made of those particles stuck on the brane, you wouldn't be able to tell that there were any extra dimensions.

The interesting thing about this is that it makes the problem of non-unique compactification worse. Now not only do you have many different geometrical ways to curl up the extra dimensions, but within those curled up extra dimensions, you can start putting in branes. You can start saying maybe we live on some of those branes. We're faced now with a bit of a conundrum. Starting from a set of unique ideas and 10 dimensions, we have a multiplicity of ways of getting down to four dimensions. The best estimates that we have right now are something like 10^{500} different possible ways to go from 10 dimensions down to four in an interesting way. By interesting, we mean a way that doesn't dissolve away very quickly—a potential place that you could imagine living in the universe of string theory.

That's the bad news. We don't like it when we lose uniqueness somehow. We're still struggling to somehow find our way through those 10^{500} different possibilities. The good news, though, is that some of those possibilities are experimentally testable. We don't know whether or not we're going to be lucky enough to be involved with one of these testable ones, but the point is that there are new ways that we realize for new dimensions of space to manifest themselves in experiments. In the old days, in the days of Kaluza and Klein, when they first suggested that there are curled up extra dimensions, all they did was just curl them up so small that you couldn't see them, by which we mean no experiment that had been done then could possibly reach them. Of course, since we haven't directly seen evidence for them, we still do the same thing today.

But, now, when you have branes, a new possibility opens up. What if we really are living on some three-dimensional brane confined in some bigger space? Remember, when we say that space is three-dimensional, we mean when we do experiments we only perceive three dimensions. For example, if you take a stick, you can take a second stick, and you can tie them so that they are perpendicular to each other. Then, you can take a third stick and you can tie that third stick so that it's perpendicular to the first two. That means there are at least three dimensions of space, macroscopically. But, try as you will, you cannot take a fourth stick and tie it so that it's perpendicular to all of the previous three all at once. It can't be done. That is an experimental demonstration that we only have three dimensions of macroscopic space; however, what if there are more dimensions, but we just can't get there? Then, the problem becomes much harder. We could have large extra dimensions and just not notice them yet.

The real issue of large extra dimensions is that even though you can confine the particles of the Standard Model of Particle Physics to a three-dimensional brane, there is one particle you cannot confine, and that's the graviton. Remember, gravity is a feature of spacetime itself. Gravity is the curvature of spacetime, so if you have some object in a set of dimensions, it's going to have a gravitational field that stretches out into all of the dimensions, not just on the brane. How do we know how many dimensions gravity feels? We have Newton's law of gravity. We have Isaac Newton's inverse-square law that says two objects pull on each other with a gravitational force that is inversely proportional to the square of the distance. The reason why it's the square of the distance—you can imagine the gravitational force lines coming from one object—they fade away as one over the distances squared because the area they are covering goes up as the distances squared. But, that's only because we live in three spatial dimensions. If we lived in four spatial dimensions, the force between two particles due to gravity would go like one over the distance cubed. If we lived in five spatial dimensions, it would go like one over the distance to the fourth, and so forth.

If we lived on a brane where the particles of the Standard Model were confined, but gravity leaked out, Isaac Newton's law of gravity wouldn't be right. We wouldn't have the inverse-square law. We would have the inverse-cube law or the inverse-fourth law or something like that. So, the clever suggestion was made in the late 1990s that maybe we do have an inverse-cube law but only on really, really tiny length scales. But, by really, really tiny I don't mean the size of a proton; I mean a millimeter across—so macroscopic length scales, the length scales where we had not yet done experiments.

So, people started doing experiments. They tried to pass Newton's inverse-square law of gravity on very, very small distances. By taking one heavy plate and bringing it very, very close to another heavy plate, they were able to squeeze the experimental limits on the size of extra dimensions down from one millimeter to a tenth of a millimeter. So, that's a whole order of magnitude, that's a lot of progress. Still, it is very plausible to us today that there are extra dimensions of space a tenth of a millimeter across; we just haven't noticed it yet.

So, we need to push forward with this kind of idea. We can't get rid of the fact that there should be a quantum theory of gravity. We have quantum mechanics. We have gravity. We have to get them together. Especially as cosmologists, we care about things like where did the universe come from? Where did the fluctuations come from that we think are due to inflation?

Our explanation of those fluctuations using inflation says that you have quantum mechanical fluctuation in the early universe during inflation, giving rise to density perturbations, and therefore we later measure the gravitational fields of those density perturbations. In other words, the inflationary scenario for explaining the observed density fluctuations in the universe involves quantum gravity in an intimate way. If we want to claim to understand those things, we're going to need to understand quantum gravity.

What we really care about, of course, is dark matter and dark energy. In the next lecture, we're going to talk about how to take string theory, with all of its extra dimensions and all of its different possibilities, and tease out implications for what the vacuum energy might be. There might also be implications for what the dark matter particle is, but right now that's much harder to get. When it comes to vacuum energy, string theory offers a very provocative—perhaps scary, perhaps even crazy—but at the very least interesting scenario for why the vacuum energy of the universe might be the dark energy we observe.

Lecture Twenty-Two
Beyond the Observable Universe

Scope:

The speed of light and the age of the observable universe are both finite. That means that we can't see the whole universe—our vision can stretch only so far. Outside what we can see, it's possible that conditions are very different; for example, the extra dimensions of string theory might be curled up in different ways, giving rise to different particles and forces. It's possible that some of the puzzles of dark energy can be resolved by imagining the universe as a huge ensemble of possibilities, within which the conditions we observe, which are hospitable for life, are atypical for the universe as a whole. This seems like a wild speculation just to explain a few numbers, but new ideas in physics make such a dramatic scenario seem plausible.

Outline

I. Cosmological horizons set a limit on what we can observe.

   A. We can't see the whole universe.

      1. There is only a finite amount of time between us and the Big Bang, which constitutes a barrier past which we can't possibly observe anything.

      2. Light, meanwhile, travels at a finite speed and, therefore, reaches back only so far before (in principle) it runs into the Big Bang. The boundary defined by the furthest point from which light can reach us is known as our *horizon*.

   B. The universe we can observe is homogeneous—pretty much the same everywhere.

      1. Is that still true beyond the horizon, potentially out to infinity? Maybe or maybe not.

      2. We have no way of ever knowing via observation; the best we can do is to contemplate different theoretical possibilities.

II. The idea that there are many regions of space with different physical properties is known as the "multiverse."

A. String theory and other models with extra dimensions suggest that our macroscopic four-dimensional spacetime can find itself in any one of a large number of possible states. In each of these states, the local "laws of physics" will appear to be different—the number of particles; their masses, spins, and charges; perhaps even the number of dimensions.

B. Inflation, meanwhile, provides a potential mechanism for turning these possible states into actual states—all at the same time but in widely separated regions of the universe.

  1. Inflation takes microscopically small patches of universe and expands them into cosmologically huge regions.

  2. It's perfectly plausible that inflation creates pockets of space in every one of the possible states suggested by string theory.

  3. In other words, we could live in one homogeneous subset of an enormously vast multiverse. We have no way of knowing for sure, but it's a plausible scenario.

III. Environmental selection is a possible explanation for the observed value of the cosmological constant.

A. "Why should I care?" is a perfectly rational response to speculations of this sort. All those other parts of the multiverse are beyond any possible observation—why should their status affect how we think in any way?

B. The answer—which may or may not be right or sensible—is that the existence of such a multiverse can change what we think of as "natural."

  1. We've already mentioned that the vacuum energy is extremely small, at least compared to its natural value. But here's the rub: If the vacuum energy actually had its natural value, we wouldn't be here to notice.

  2. Life could not exist if the cosmological constant were very large.

    a. If it were large and positive, it would accelerate the universe so fast that atoms and other structures couldn't form.

    b. If it were large and negative, the universe would recollapse in a microscopically short period of time.

3. Even though the observed vacuum energy seems unnatural, in a sense, it couldn't be very different; if it were, we wouldn't be measuring it.

C. Thus, the concept of a multiverse leads us to think in terms of *environmental selection*: In a collection of many different environments, we will only ever observe those that are compatible with our existence.

1. If such an ensemble really does exist, it shouldn't be at all surprising that the vacuum energy is small.

2. We could imagine asking, right here in our Solar System, why we were so lucky to have life arise on the hospitable Earth rather than the overheated surface of the Sun, which after all, is much larger. But of course, it's not luck at all; life can't arise on the surface of the Sun, so that was never an option.

D. What does this mean for dark matter and dark energy? For dark matter, not much: The existence of dark matter doesn't seem to be crucial for our way of life. But dark energy is a different story.

1. If the dark energy were large and positive, the acceleration would be so rapid that atoms couldn't form, much less galaxies. And if it were large and negative, the universe would recollapse in a fraction of a second.

2. Even though the observed magnitude of the dark energy seems unnatural to us, it had to be close to the current value for us to be here in the first place.

3. If, indeed, we live in a multiverse, it's not surprising that we live in a part of it that is hospitable to us being here.

E. Such reasoning often goes under the name of the *anthropic principle*.

1. Many people object to this principle on philosophical grounds, especially because the existence of the multiverse is impossible to verify (or disprove).

2. It's important to realize, however, that the multiverse is not by itself a scientific theory; it is, at best, one component of a larger theory.

3. The ultimate test is whether or not that theory (which is, at present, completely hypothetical in its own right but should, eventually, include string theory, as well as inflation) makes other predictions that are testable against experiment. If not, it can't really qualify as science.

Recommended Reading:

Livio, *The Accelerating Universe*, chapters 9–10.

Randall, *Warped Passages*, chapters 19–20.

Vilenkin, *Many Worlds in One*, chapters 10–15.

Questions to Consider:

1. Should physics ever address parts of the universe we can never observe? Does this count as "science" in your estimation?

2. Imagine that researchers managed to use environmental/anthropic reasoning to make an explicit prediction about particle physics, which was later verified at the Large Hadron Collider. Would that count as evidence in favor of this approach?

3. We don't know what forms intelligent life could possibly take in different physical conditions (or even in the conditions we observe). What do you think are the minimal requirements for the laws of physics and conditions in the universe to support life?

Lecture Twenty-Two—Transcript
Beyond the Observable Universe

In this lecture, we're going to get the payoff from the previous two lectures on inflation and string theory. When we talked about string theory, we realized that the thing that we would like to be the case—namely, that what we think is the unique string theory that lives in higher numbers of dimensions predicts things like what is the dark energy and what is the dark matter—doesn't quite turn out to be true. String theory predicts too many things. As far as we can tell, it is compatible with all sorts of different possibilities. Nevertheless, it has an impact on how we think about dark matter and especially about dark energy.

What we'll talk about in this lecture is how ideas from string theory and inflation changed our notion of what constitutes a natural value for the vacuum energy. Remember, after we talked about dark energy existing, and going through all the different possibilities for what it might be, the possibilities of quintessence and changing gravity were interesting, but ran into problems with experiments. The possibility that it is vacuum energy, an absolutely constant amount of energy in every cubic centimeter of space, didn't run into any problems with experiments, but seemed very, very unnatural. The value that vacuum energy would have to have is just very, very different from the value that it seems to actually have in our universe.

So, string theory has the chance to recalibrate our notion of what it means for a number to be "natural," and this is a long process that we're going to go through in this lecture. The punch line is that string theory says there can be many different phases. All the different ways of taking the extra dimensions of string theory and all the different branes and other things it predicts, and compactifying them down to get a four-dimensional spacetime like the one where we live, can correspond at low energies to different phases of spacetime.

Just as water can come in different phases of liquid water, solid water of ice, and water vapor, it's the same underlying thing. It's not like there are three different theories—a theory of water vapor, a theory of ice, and a theory of liquid water; it's the same stuff manifesting itself in different forms. Likewise, what string theory is saying is that there are perhaps 10^{500} different phases of spacetime—

all of these different ways in which the fundamental vibrational modes of the string can show up as particles, as numbers of particles, and as the vacuum energy. The vacuum energy is going to be a number that changes from phase to phase, from compactification to compactification. It will take on all sorts of different values—10^{500} different possible values. Within that ensemble, we're going to get a lot of different possibilities. One of them might be the one in which we live. This idea is sometimes called the *anthropic principle*, the idea that we're picking out among a huge ensemble of possibilities those possibilities that allow us to exist. The reason why that's a sticky situation is because we don't know what "us" means in that sentence. What do you mean "us?" What kind of definition do you have for what counts as intelligent life? So, we're not going to go into any of those issues in any detail. What we're going to rather think about is in terms of *environmental selection*.

It's not a surprise that if you live in a universe, or in a set of universes, in which conditions can be very different that we're going to observe those conditions that are hospitable to us living there. That is just a tautology; that is not surprising. The interesting part is when we go from the tautology to using it to make predictions. Within this ensemble, what is likely for us to observe? We might be able to change our notion of what you would expect ahead of time by realizing we live in an ensemble of many possible universes rather than in one unique thing.

We live in such a universe that we can't see the whole thing. We live in a universe where the observable part is defined by a *horizon*. We send back light rays into the past and because the speed of light is finite, those light rays hit the Big Bang, hit a boundary at a finite distance. There are almost certainly parts of our universe which we can't see because they are simply too far away. That is not a surprising claim or controversial in any way. We can certainly see out to the cosmic microwave background. If we're trying to be clever about it and we learn how to use neutrinos or something like that, we can push that back a little bit further, but there is still a very clear demarcation past which we can't possibly see, which is given to us by the Big Bang itself. What is beyond that part that we can see?

The part we can see seems to be homogeneous and isotropic. Not only is the configuration of stuff more or less the same, the same density of stuff from place to place, but it seems from our observations as if the laws of physics are the same. It's not true that the charge of the electron has a different value in one part of the universe than in the other one. As far as we can tell, they are the same everywhere. Is it possible that you can just extend that understanding infinitely far? Is it conceivable that we live in a universe where conditions really are the same everywhere, even outside what we see? The answer is absolutely yes. There is no reason we can give—either logically or within the laws of physics, as we currently understand them—against the idea that the universe is truly the same everywhere, even outside what we observe.

But, by exactly the same criteria, there is no possible reason we can currently give to say that the universe is the same everywhere. It is absolutely just as reasonable to say the universe is very, very different outside. On the one hand, you might say that it is very parochial, very anthropocentric of us to take our local universe and extend it all over the place. On the other hand, you might say that it's not very parsimonious to have a universe that is wildly different. We have a universe that looks very nice as it is. Why not just take the simplest possibility and extend it all over the place?

The point is that we can't answer this question just by pure thought. When we're in a situation like that, what we have to do is to allow for both possibilities. We don't need to make a decision. We're going to ask what happens if the different things are possibly true. For this lecture, we're going to ask what happens if there are different regions of the universe where conditions are very different? We're going to call this the "multiverse," but it's not a multiverse in any metaphysical sense. It's not like there are different universes that are separate from each other by some profound difference. There are just different regions of space into which we cannot get.

That's an interesting thing to think about as a possibility, but it's string theory and inflation that takes this possibility and makes it very tangible. In other words, string theory plus inflation gives us a set of ideas from which we can talk about the possibility of a multiverse in a scientific way. It is string theory that allows space to take on different conditions, not just different densities, but different phases. The different ways we have of taking the extra dimensions of space in string theory and curling them up give us different low energy physics.

In our current world, we seem to have four dimensions of spacetime, so three of those dimensions are large dimensions of space, and we have the Standard Model of Particle Physics. The Standard Model is characterized by a set of particles and also by a set of numbers, so we say we have certain fermions, we have certain bosons, etc., certain interactions relating the particles, and also certain parameters. Those parameters are the charge of the electron, the mass of the electron, the mass of the up quark, and so forth. One of those parameters is the vacuum energy.

When string theorists realized that there was more than one way to compactify the extra dimensions, they realized that in fact if there was going to be more than one way, there was going to be a huge number of ways—something like 10^{500} is the current best guess. To this set of possibilities, they've given the name the *landscape*. The string theory landscape is something you can think of as like some jagged landscape here on earth, and every little minimum, every little local valley, is a different place you can live. Different valleys, of course, would have different local conditions—different temperatures, different densities, and different heights. That's an analogy to what we have in string theory where there are different ways to curl up the extra dimensions, perhaps not an infinite number of ways, perhaps it's not anything goes, but the number of ways we can get is still very large; 10^{500} (1 followed by 500 zeros) is a huge number of possibilities.

I should say that we're very far from being certain that this is true. Currently, our best understanding of string theory says that there are perhaps 10^{500} different ways to curl up the extra dimensions; however, it is certainly conceivable that the current state of the art just isn't good enough to say that for sure. In other words, it's possible that once we understand string theory better we will come to understand that even though you can curl up things in different ways, they don't stay curled up like that. You might curl up the extra dimensions of a certain configuration, but they quickly unwind into a different one. It is, therefore, still quite possible that once we understand string theory better than we do today, we will narrow it down to a very small, perhaps even unique, set of possibilities. That is a goal to keep in mind, if it's true, but the current best guess on the basis of what we seem to understand right now is that these different phases of string theory really are stable and really can exist in principle.

What inflation does is take phases of string theory that can exist in principle and gives them a way to actually exist in practice. What inflation says is that at very early times in the history of the universe, we don't know exactly what was happening, but perhaps there were chaotic fluctuations. Different conditions were going on all over the place; it was very high temperature, with very large fluctuations from place to place. In some tiny little patch of that initially chaotic system, you've got a domination by something like dark energy, some inflaton field with an approximately constant energy density. That inflaton caused that little patch to accelerate, to expand to a huge size. Eventually, the energy in the inflaton reheated into matter and radiation, and we see what we see today. So, if you take the idea of inflation and combine it with the theory of 10^{500} different possible stable final states for the compactified dimensions of string theory, what you get is a way to make those possibilities real. By starting inflation in slightly different conditions, by allowing inflation to go on in slightly different ways, you can have different trajectories, all of which populate any one of the 10^{500} different valleys in the landscape. In other words, we're imagining a multiverse that starts out in some chaotic condition. Then, through inflation happening in different parts, in different ways, using different physics, we get a final condition in which you get huge bubbles of universe, all of which could be in any one of the 10^{500} different phases of string theory.

Clearly, there are a lot of details to be filled in when we talk about something like that. Right now there is a lot of hand waving involved. We don't know the correct picture. But, that kind of picture is perfectly plausible. It might very well be, according to what we understand right now, that the universe we observe is a tiny infinitesimal fraction of everything there is. It is arguable that that is the lesson of Copernicus. If you're not putting us at the center of the universe, you shouldn't assume that the conditions that we observe are the same conditions that obtain all over the universe.

One thing that I want to talk about is whether or not this is legitimate, whether or not it's OK to talk about regions of the universe, which we cannot see. People will say if you can't observe these different regions of the universe, they have no effect on local physics. They have no effect on what is going on in your region of the universe and they never will; therefore, talking about them isn't even science. Who cares about whether or not they are there? The reason why we would possibly care—and we'll talk about this in more detail later—is because living in an ensemble changes our notion of what is natural. If you only live in a unique universe, then you might guess that the constants of nature will naturally take on the values that would seem easy to us if everything was of order one, if there were no large differences between the different numbers that you saw. But, if you live in an ensemble, then every possibility happens, even some extremely, extremely rare ones. If one of the extremely rare possibilities is somehow more hospitable for us to living there, then we should not be surprised if we live in one of the rare possibilities. It would be very natural for us to live in a universe in which the parameters of nature somehow didn't seem natural. That's why it's worth thinking about this possibility that we live in an ensemble, a multiverse of different phases of string theory.

Other people will say, in a related way, that it's not a scientific theory if it doesn't make very particular predictions, but I want to point out that the multiverse is not strictly speaking a theory. The multiverse, if it's there, is a prediction of a theory. The theory is string theory combined with inflation. That's not a very exact theory right now. We don't understand either inflation or string theory well enough to tell you precisely what the predictions are. But, the goal of this kind of way of thinking is the following. Someday we will be able to do experiments that convince us that a certain theory of quantum gravity is correct. Hopefully, we will be able to narrow down, on the basis of data, on the basis of experiments, which version of string theory, if any, or which version of quantum gravity, correctly describes our world. We will, within those experimental constraints, be able to say the following fields can act like inflatons, can make the universe expand. In other words, on the basis of data, we will build a framework, which makes a specific prediction for what the multiverse should be like. Within that prediction, we can begin to make sense of questions like: What should people observe who live in that ensemble? Even though the prediction of the multiverse is itself not testable, it might be an airtight prediction of a model that has other testable predictions. That may or may not be a utopian goal, but that is the kind of thing that we're shooting for when we think about these ideas in the back of our head.

OK, so now the philosophy is a little bit out of the way. Let's try to talk about putting this to work. What if we really do live in a multiverse? What if there really is an ensemble of different places that we don't observe where conditions are very, very different? It goes without saying that within that ensemble of different possibilities, there will be very strong selection effects when you ask, what is observed by a typical member of that ensemble? We imagine, roughly speaking, that some of these places in the multiverse are very inhospitable. We just can't live there. So, it is not surprising that no one is observing them if observers cannot exist. Now we admit that there is a very, very big question here of what is an observer? What is somehow what people call a conscious person or some intelligent scientist who can live if the laws of physics are very, very different? You could ask very detailed questions about this.

For example, imagine a universe that was almost exactly like our own, but in which the proton was a little bit heavier than the neutron. This is something that we talked about when we talked about the Standard Model of Particle Physics lectures ago. What would happen is a proton would then decay into a neutron. Heavy particles decay into light ones. Instead of a world made of atoms where you have atomic nuclei surrounded by electrons, you have a world made of neutrons. It should be clear that a world made of neutrons is very different than the world in which we live. We can't have chemistry in a world made of neutrons.

You might therefore ask, can you have intelligent observers in a world made of neutrons? Some people think they know the answer to that question. I do not think that I know the answer to the question can you have intelligent observers made of nothing but neutrons? I could hypothetically conceive that neutrons could get together. You could have little nuclei made of one, or two, or three neutrons; these nuclei could even get together to make neutron molecules and those molecules could build up into neutron amino acids. I truly don't know whether or not there is sufficient room for complexity in a universe made almost all of neutrons to support intelligent life; therefore, I don't care. Therefore, I'm not going to talk about any predictions you can make at a very tightly quantitative level that say if we increase the mass of the proton by 10 percent then life cannot exist. We're going to keep an open mind about that. We're going to stick to things which, I think every reasonable person would agree on, do say something about whether or not conscious observers can exist.

In particular, we're going to talk about the vacuum energy. The vacuum energy is one thing that we think has a very, very unnatural value. The vacuum energy, if it is the dark energy, is 10^{-120} times what we would have thought would be its natural value. That seems preposterously finely tuned from the point of view of ordinary particle physics. If everything were natural in the Standard Model, the vacuum energy would be at the Planck scale. It would be 10^{120} times bigger than it is today; however, nobody I've ever met claims that life could exist if the vacuum energy were that big. If it were that big and a positive number, then the acceleration of space would make it absolutely impossible to form planets or, for that matter, individual atoms. You couldn't even make a proton in a universe where the vacuum energy was at the Planck scale. Everything would be ripped apart very, very quickly. You would be left with an empty universe almost instantaneously. There is no room to form life in such a universe. If the vacuum energy had the magnitude of the Planck scale, but were negative it would make the universe recollapse in one Planck time—much too small of a time scale to have any realistic or interesting particle physics, much less actually form life.

In other words, when it comes to the cosmological constant, we have a very strange situation. We have a natural value for it to have. The cosmological constant is just a different word for the vacuum energy and it should be at the Planck scale. That's its natural value. But, if it had its natural value, we would not be here to talk about it. On the one hand, it's not surprising at all that the cosmological constant doesn't have its natural value. It can't have its natural value or we wouldn't be here thinking about it. The question is why do we live in a universe in which the vacuum energy, the cosmological constant, is small enough to allow us to be here? There are basically two different possibilities. One possibility is we just got lucky. In other words, there is no reason why the vacuum energy is small enough to allow for the existence of life. It just happened to be that.

Within that possibility—the just-got-lucky idea—there are two sub-possibilities. One says we truly just got lucky. It's not only that the vacuum energy is small, but it's just a random number. There is no dynamical physical explanation for why it's small. It was just randomly chosen and, happily, it turned out to be 10^{-120}, what it should have been; therefore, we can exist. That is an absolutely possible theory of the value of the cosmological constant. I can't tell you that theory is wrong. But, for obvious reasons, it's kind of unsatisfying. That would be a truly lucky roll of the dice for us for the vacuum energy to be that small. The other possibility within the idea that we just got lucky is that there is a dynamical mechanism. The thing that we got lucky about is not that the cosmological constant is a random number that is small, but that the cosmological constant is small because there is some physics that we haven't yet figured out that makes it small. We're lucky for the existence of that physics, not for the random throw of the dice that makes the vacuum energy very small. But, right now, we don't know what that physics is. We're just guessing at it. Many people are hopeful that we'll find it at some point, but we just don't know.

The other possibility besides that we just got lucky that the cosmological constant is small is that we had to get lucky in the sense that the cosmological constant is not a once and for all constant of nature. It's an environmental variable that takes different values in different places. If the cosmological constant takes different values in different regions of the multiverse, and in some of those regions it's small enough that life can exist, then it's not a surprise. Then, it's not that we got lucky that we exist in those regions.

Let me give you an example. Imagine that, as an analogy, there were astronomers who lived on a planet where the atmosphere was opaque. You could never see the sky. On this planet, the temperature was very mild and never changed. It was 70 degrees every day, but you couldn't see anything besides the clouds overhead. In such a universe, in such a hypothetical situation, what would the scientists who live on that planet try to do? They would try to understand the temperature that was on their planet. They would say: Are there laws of physics that predict that the temperature will always be 70 degrees and we can be here? Other physicists on that planet would say, my idea is that there are also other planets, and on those other planets, the temperature is very different. But, there are just so many planets that life is going to arise on the planets where the temperature is nice. Other people will say, oh, come on, that's just philosophy. That's not science. We can't see these planets. How can we be talking about them?

©2007 The Teaching Company.

Our current situation is actually much like that. Just like there are clouds in the sky in this hypothetical planet, we have a horizon given by the microwave background, past which we can't see. Maybe there are other regions of the universe out there where conditions are very different. The fundamental question before us now is: Is the vacuum energy a once and for all constant of nature for which there should be some equation that predicts its value, or is it an environmental variable? Is it different from place to place, and therefore chosen by a selection effect, rather than a deep law of physics?

If we were going to decide between those possibilities, it would be nice to take this idea that there really is an ensemble of different possibilities, and actually use it to make some sort of predictions. To not just use it to make us feel good about the value of the vacuum energy, but actually do something precise with it. What you want to do is to consider the entire ensemble, and within that ensemble ask, what does a typical observer living in that ensemble of the multiverse actually observe?

This exercise was undertaken in the 1980s by Steven Weinberg, a well-known physicist who actually won the Nobel Prize for his work on the Standard Model of Particle Physics on the W and Z bosons. What Weinberg says is that we can actually attach numbers to the statement. If the cosmological constant were large and positive, it would rip things apart. If it were large and negative, it would make the universe recollapse. He did a little calculation and he argued that the typical value of the vacuum energy that would be observed by a conscious observer is something like 10 times as big as the matter density, by which he means the vacuum energy could be anywhere from -10 times the matter density to +10 times the matter density. So, you pick a random number between -10 and 10. A typical random number between -10 and 10 is not going to be 10^{-5}. It's going to be of order one. It's going to be -3 or +5 or something like that.

Weinberg, in 1988, made a prediction that someday we would observe a non-zero cosmological constant. That was a prediction that he made before we went out and found it. Then, 10 years later, we did find it. In fact, it was consistent with his prediction. The vacuum energy we think we have observed is two or three times the matter density. That is consistent with being between -10 and 10. Not only is it between there, but it's a kind of a typical number you'd expect it to be if you picked a random number between -10 and 10.

What does that mean? It means that you can predict something, at least you can try. You might want to predict other things. For example, you might try to predict the masses of the elementary particles—the mass of the electron, the mass of the quark, and so forth. Even better, you'd like to predict something we haven't yet observed to make sure we're still on the right track. The problem is the environmental selection that says we can only exist where conditions allow for us to exist isn't a very strong constraint on things we haven't yet observed.

What is the dark matter? Is the dark matter a WIMP? Is it the lightest supersymmetric particle? Or, is it an axion? The environmental selection principle doesn't tell us which one. We can exist just as well with both axionic dark matter or supersymmetric dark matter. You could even ask, how much dark matter should there be? Some brave souls have tried to argue that the environmental selection principle predicts there should be dark matter. The reason why is because the more dark matter you have, the more structure you form in the universe, the more galaxies, and therefore the more observers. To me, personally, this seems like a stretch. This seems more like a post-diction than a prediction. We already know there is dark matter there. We're trying to justify it after the fact. But, I think that for vacuum energy, there is some possibility of something going on there. We don't know exactly what will go on. We don't yet know how to predict things we haven't observed, but it's still interesting to try to push our current knowledge of how the ensemble would work past what we get.

Nevertheless, let me give equal time to the objections to this way of thinking. I should say that currently within the physics community there is a sharp divide. There is a set of people who take very, very seriously the idea of the multiverse as an explanation for the observed vacuum energy. There is an equal number of people, maybe even a larger number of people, who think it's completely crazy. I'm going to give you the arguments from the people who think that even talking about this is not what we should be doing.

There are two basic kinds of objections to talking about the multiverse, what I will call the grumpy old man objections and what I will call the sensible objections. The grumpy old man objections are the ones that just say, you know, doing stuff like this isn't science. You can never observe the stuff out there. You're giving up on our attempts to understand the universe based on evidence, and observation, and experiment. These are unconvincing to me as reasons to not think about the multiverse. First, because it might be like that. It might be the case that outside what we can observe the universe takes on very different conditions. Whether or not we can observe them, it might be the truth. Ultimately, our goal is to get at the truth by whatever method we can. If the way that we get there is by naturalness arguments rather than direct experiments, then that's what we have to live with.

The second is that the multiverse is not, as I said before, a theory that makes strong predictions; it's a prediction of a theory that we can use to recalibrate our notions of naturalness. The point is that when we look at naturalness, when we look at quantities we measure in nature and say that one doesn't look right, that looks unusual to us, we're taking that as a clue. We're saying that we really don't understand the final laws of physics, but we're trying. Sometimes, the way that we are moving toward the final laws is to do an experiment that gives us more information. But, other times, the way that we move toward a better understanding of the final laws of physics is to look at unnatural features of our current understanding.

For example, that's what happened with the horizon and flatness problem in inflation. We thought that there was a good understanding of how the early universe behaved, but we didn't know why. Thinking about the reasons why the universe would be flat and nearly homogeneous led us to inflation. Thinking about the value of the vacuum energy might very well make more sense in the context of a multiverse that might lead to a better understanding of other things.

The non-grumpy objections I take more seriously. There is a set of objections to thinking about the multiverse that I think are quite reasonable and we should take seriously. Basically, what these objections are saying is that even if in principle we do live in a multiverse with various different things going on, as a matter of practice, it is impossible to extract from that any detailed predictions. Against this, you could say, well, Steven Weinberg who won the Nobel Prize made a detailed prediction. What about that? But, there is a good objection to the fact that Weinberg made a prediction and sort of got lucky. The point is he made a very specific calculation that did the following thing. He said: What if we had an ensemble of the multiverse in which conditions in all the different parts were exactly the same as they are here except for the cosmological constant? He only allowed the vacuum energy to change and he derived a prediction of the vacuum energy to be pretty close to the matter density. But, that's not what the actual multiverse situation tells us to do. It says that, from place to place, everything changes—not just the vacuum energy, but the set of particles we have, the way that inflation works, and so forth. If you were allowed to change not only the value of the vacuum energy, but other quantities, like the value of the dark energy or the amplitude of the initial density perturbations, you would make a much less strong prediction. The true multiverse prediction for the value of the cosmological constant is not nearly as precise as Weinberg's calculation makes it seem to be.

The truth is we just don't understand the multiverse well enough right now to make detailed predictions using it. What we're doing is saying that there is this set of universes, set of places in the universe, where conditions are different. There could be an infinite number of such places. There could be an infinite number of observers in every one of those places. We're trying to say what does a typical observer in that ensemble actually observe? The state of the art is nowhere near right now being able to answer that question. It may be because we don't yet understand inflation and string theory very well, or it may be because we never will; there isn't any right answer to this question. If there are an infinite number of observers who observe this, and an even bigger infinite number who observe something else, how are we possibly to say which is more likely for us to be there? This is all speculation at this point in time, but the reason why it's worth going over is because it's speculation that might turn out to be right.

The real lesson is that we don't have any convincing theory of the vacuum energy, and we're driven to the environmental selection in the multiverse as the best we can think of right now. It might turn out to be right. The worst version of the anthropic principle would be to think that the universe arranges itself into the way that you think it should be. What we should do is to keep an open mind about all the possibilities, as we get more and more data to help us zoom in on which one is ultimately correct.

Lecture Twenty-Three
Future Experiments

Scope:

The concordance cosmology, with dark matter and dark energy, fits the data very well, but we would like to have a better fundamental understanding of these dark substances, and the way to achieve this is through new experiments. Astronomers are planning a suite of new observatories on the ground and in space to better probe the acceleration of the universe, the evolution of cosmic structure, and the dynamics of galaxies and clusters. Physicists are also looking forward to new results from particle accelerators, underground detectors, and tabletop experiments that will probe the nature of gravity on small scales.

Outline

I. We have a model that fits the data extremely well.

   A. The current best-fit model of the universe invokes a number of amazing ideas, such as dark matter, dark energy, and density fluctuations that may be generated by a period of inflation in the very early universe. Nevertheless, the model fits an impressively diverse collection of data. So far, nothing we see about the universe is in substantial conflict with this standard cosmological model.

   B. However, we want to do better. We're not content to *describe* the universe; we want to *understand* it. To that end, of course, we need better theoretical ideas, and the easiest way to hit on those ideas is to have new experimental input.

II. There are many ways to look at the universe with photons.

   A. Since ancient times, we've gathered information about the heavens by looking with our eyes, suitably aided by technology.

   B. In this day and age, we can detect photons across the electromagnetic spectrum, from long-wavelength radio waves to short-wavelength X-rays and gamma-rays.

1. With this technology, plans are on the drawing board to dramatically improve a number of different kinds of cosmological observations.
2. Such efforts include the following:
   a. Deeper probes into what we can learn from the cosmic microwave background, e.g., comparing the fluctuations at large angles in the sky to fluctuations at smaller angles in order to measure the amounts of (and fluctuations in) dark energy, dark matter, and ordinary matter in the early universe.
   b. Surveys of galaxies and large-scale structure, e.g., the Sloan Digital Sky Survey (Arizona) and the Two-degree Field Survey (Australia).
   c. Maps of gravitational lensing to infer the location of dark matter and the evolution of structure.
   d. Detailed examination of clusters of galaxies, because the number of clusters of galaxies in the universe at different distances (at different redshifts) is a very promising way of figuring out how much dark energy there is.
   e. More measurements of Type Ia supernovae at large distances in order to figure out whether the dark energy is truly constant or slowly varying.
   f. Gamma-ray satellites to look for annihilating dark matter, e.g., GLAST (Gamma-ray Large Area Space Telescope).

III. Increasingly, we can also use particles besides photons.

A. It's only relatively recently that technology has developed to the point where we are not limited to using only electromagnetic radiation to probe the universe.

B. Current and upcoming observations will use:
   1. Cosmic rays, which are high-energy protons and nuclei from cosmological distances.
   2. Neutrinos; those currently measured are only from the Sun, but we're looking for other sources.
   3. Gravitational waves; they have not yet been directly detected, but more advanced observatories have recently become operational.

a. In 2007, LIGO (the Laser Interferometer Gravitational-Wave Observatory) began to collect data at the rate for which it was designed.

b. Even more exciting are plans for a gravitational wave observatory in space, known as LISA (the Laser Interferometric Space Antenna).

C. Direct detection of dark matter particles is an exciting future possibility, e.g. CDMS (Cryogenic Dark Matter Search) in Minnesota and DAMA (at the Gran Sasso National Laboratories) in Italy.

IV. A great deal of information relevant to cosmology comes from experiments performed here on Earth.

A. One of the deep lessons of modern physics and cosmology is that we can learn a lot about the universe by doing experiments close to home.

B. Ongoing projects include:

1. Solar System tests of gravity, especially using satellites traveling to other planets, e.g., the STEP Mission (Satellite Test of the Equivalence Principle).

2. Tests of gravity on very small scales—less than 1 millimeter.

3. Searches for new weak forces stretching over macroscopic distances.

V. Most promising of all is the *Large Hadron Collider* (LHC).

A. Becoming operational for the first time in 2007, this is a new machine at the CERN laboratory in Geneva, Switzerland, operating at the high-energy frontier of particle accelerators. It will peer into the regime where we might expect to see the Higgs boson, dark matter candidates, and evidence for or against supersymmetry.

B. Chances are very good that discoveries from the LHC will dramatically alter our ideas about particles and forces and how ordinary matter fits in with dark matter and dark energy.

Recommended Reading:

As with Lecture Nineteen, the projects described in this lecture are all very new and cutting-edge (as of this writing). Books appear far too slowly to keep up to date with this type of experimental progress. Stay tuned to such journals as *Scientific American* and *Sky & Telescope* for news about the latest progress in experimental physics and observational astronomy.

Questions to Consider:

1. Experiments cost money. How should we balance our interests in scientific discoveries against other goals, such as the human exploration of space or the solutions to problems here on Earth?

2. Can you think of any other interesting kinds of experiments, other than the ones described here, that would be relevant to the questions of dark matter and dark energy?

Lecture Twenty-Three—Transcript
Future Experiments

We've set ourselves the task of understanding the dark side of the universe—the dark sector that makes up 95 percent of the stuff that we seem to have in our contemporary cosmology. For the last few lectures, our strategy toward understanding the dark sector has been to think really hard, to think about where the universe came from and how inflation could have taken us from a tiny patch to the big universe we see, and to think about how the fundamental laws of physics work in terms of the possible role of string theory, extra dimensions, and the multiverse.

These ideas come into our conception of dark matter and dark energy because they give us a framework in which to think about what are natural values for different things to have—what is the natural value for the dark energy to be, for the dark matter to be, etc. But, thinking hard is great; sometimes it's all you have. But, experimental data is always better. In either case, what we're trying to do is to go beyond our current understanding. Sometimes you go beyond your current understanding because your current understanding is mutually inconsistent with itself. Special relativity and Newtonian gravity were inconsistent. Einstein used that to invent general relativity. Now we have quantum mechanics and general relativity being inconsistent. That's how we get to string theory. But, it's never as good as actually having a piece of data that says your current theory is wrong. Part of the reason why we have not had great theoretical progress in particle physics in the last 30 years is because the Standard Model is too good. It fits all of the data really, really well. To move beyond, to get a better idea for what's going on next, it is always better to get a new experimental piece of information that doesn't fit our current theory.

The good news is we have a huge amount of experiments coming on line in the next few years, which hopefully will teach us more beyond what we already know. I will remind you—you've probably heard it by now—we have a model that fits the data. We have a picture of the universe in which 5 percent of the universe is ordinary matter, particles from the Standard Model of Particle Physics, 25 percent is dark matter, and 70 percent is dark energy. The dark matter is some particle that is massive, moves slowly, and doesn't interact very strongly. The dark energy is some form of energy that is smoothly spread through space and slowly evolving through time.

That's fine. That's a model that fits the data. It fits all the data we have. Should we therefore declare victory? If we stop collecting more data, we wouldn't have any discrepancies. But, we want more than that. We want to understand what is going on. Even though we know there is dark matter and we know there is dark energy, we don't know what these things are, so we're going to push our experiments forward. Remember that when we push these experiments forward, sometimes we learn a little bit more about what we were studying, but sometimes we were completely surprised. One of the supernova groups that tried to measure the deceleration of the universe titled themselves "Measuring the Deceleration." They eventually discovered the acceleration. Those are the best kinds of experiments. Even though we can talk about what experiments we'll be doing over the next few years, we cannot talk about what we will learn from them. We have things we hope to learn, but we need to keep an open mind. Maybe we will be surprised.

We can characterize the kinds of experiments we'll be doing by the kinds of things they will be looking at. In other words, are they looking at photons, are they looking at neutrinos, or what have you? Photons, of course, are our favorite way to look at the universe. They always have been. We'll start by looking at what we can learn by measuring photons coming from the sky. There are also photons we can make here on the ground, of course. Traditionally, all of astrophysics and all of astronomy meant studying photons we get from the sky—photons across the electromagnetic spectrum from radio waves at very, very long wavelengths, to microwaves and infrared, up to visible photons that we see with our eyeballs, and then to shorter and shorter wavelengths, through ultraviolet X-rays, and ultimately gamma rays. It says something about the state of our progress that we even make a meaningful distinction between looking at the sky in photons versus anything else. For the first time in human history, over the last couple of decades we've begun to look at the sky in other kinds of things. Nevertheless, it's still photons that are giving us the overwhelming amount of information.

You might think that we could organize the kinds of experiments we're going to be doing by what kinds of photons we'll be looking at—X-rays versus ultraviolet, for example. However, it turns out that for some kinds of objects in the sky, they are best understood by looking at the same object in different wavelengths; therefore, let's organize how we will be looking at the universe by what we will be looking at rather than what wavelength of photon we'll be using. Let's go from the largest things in the sky to the smallest things. The largest thing is really the cosmic microwave background. The microwave background is the leftover radiation from the Big Bang, a snapshot of what the universe looked like about 400,000 years after it began. It's the tiny variations in temperature that the microwave background has from place to place in the sky that teach us a tremendous amount about what the early universe was really like.

It's already been a treasure trove of information about what the universe was like then and what it must be like now. We've learned from the temperature fluctuations in the microwave background, number one, that space is flat; therefore, you need—if the Friedmann equation is right—enough energy density to account for a spatially flat universe. Number two: Ordinary matter is not enough. In order to explain the different sized splotches of hot and cold that we see in the cosmic microwave background, it is necessary to introduce stuff other than ordinary matter, and that of course is the dark matter.

What will we learn by going beyond what we've already done with the microwave background? The first thing we can do is the most obvious, which is we can have better resolution. We can look at smaller angular scales. The most interesting angular scale, if you had to pick one on the microwave background, is an angular scale of about one degree. That one we've pretty much mapped out. But, there could be surprises waiting for us at smaller and smaller scales.

One of the predictions of inflationary cosmology is not just that there are perturbations at early times, but relationships between the amount of perturbations at different wavelengths, at different distances in the sky. So, a big project for the future will be comparing the fluctuations at large angles in the sky to what fluctuations look like at smaller angles. To do that, we have a satellite that is supposed to be launched in 2008 called the Planck Surveyor, after the German astrophysicist, Max Planck. That's going to go up and give unprecedented accurate views of what the microwave background sky looks like at very, very tiny angles. What you will get from that is a high precision way of measuring the cosmological parameters, not just how much dark energy there is, but how much dark matter there is, how much ordinary matter there is, and what kind of fluctuations there were in the very early universe that grew into the galaxies and structures we observe today. But, even then, we won't be done with the microwave background. We're still going to try to squeeze more information out of it. We've already mentioned that we have recently discovered the polarization of the microwave background. The wavelength that we're getting from the CMB are polarized in different directions depending on the temperature fluctuation.

But, in addition to the intrinsic polarization, which we've already observed, there is an additional component, hypothetically, due to gravitational waves created by inflation. These have not yet been observed. One of the major projects upcoming someday, hypothetically with a satellite mission, will be dedicated to looking for the polarization imprint of gravitational waves from inflation. If you could find those and they had the properties that were predicted by our currently favorite models of inflation, that would be the best evidence we know how to get that something like inflation was true in the early universe.

Right now, we have models of inflation that are consistent with everything we observe, but their predictions are kind of vanilla. They are that the universe should be flat, that their perturbations should be approximately the same on all scales. The tensor perturbations, which are what we call the perturbations we get from gravitational waves, are a specific and quantitative prediction from inflation. So, if that turned out to be right, we would really think we were on the right track.

After the microwave background, the next largest thing we can look at is large-scale structure in the universe. In other words, we can map out the stuff that we see in the later universe—the galaxies and the clouds of gas and dust. Right now, the current state of the art is something called the Sloan Digital Sky Survey along with a slightly smaller survey called the 2-Degree Field. The Sloan Survey has about a million different redshifts for galaxies collected at a telescope in Arizona. The 2-Degree Field Survey has about 250,000 redshifts collected from Australia. Together, they have given us unprecedented views of the way in which structure stretches across the universe.

So, by matching to the microwave background, which is a snapshot of the universe 400,000 years after the Big Bang, to the redshift surveys, which are a snapshot of the universe 14 billion years after the Big Bang, we're beginning to piece together how structure has evolved from early times to late times. That is a messy and complicated astrophysical problem. Even if you knew from inflation what the initial perturbations were and you knew the ingredients of the universe in terms of dark matter and dark energy, it's still quite a bit of work to trace from those initial perturbations to what we see today. There is a lot of dirty important astrophysics that needs to be done and observations of the distribution of structure are going to be crucial in making sure that we're on the right track when doing that.

One of the most important roles of redshift surveys, of surveys of galaxies across the sky, will not simply be to use those data all by themselves, but to be to compare them to other kinds of data. One thing we will be doing is surveys of gravitational lensing. Only very recently have we been able to develop cameras that can take images of what's in the sky to such precision that you can tease out the slight perturbations of the shapes of galaxies due to gravitational lensing. It's this weak gravitational lensing that is going to be a treasure trove of information for future cosmologists. Remember that if gravitational lensing happens, the beam of light that you get from a background galaxy is deflected by some mass distribution and by the amount of deflection you can measure how much mass there is. That's more or less the simple basic picture, if you're lucky enough to get one heavy object, acting as a lens, and one background object that you observe on the sky.

The real universe, of course, has many, many galaxies in it and many, many clusters of galaxies. It's a messy configuration. That's bad if you only want to take one picture and get one piece of information, but what we're learning how to do is to use that messiness to our advantage. We are now learning how to make 3D maps of where the stuff is in the universe using the amount of gravitational lensing that is undergone by galaxies at different redshifts. By combining redshifts surveys to tell us where the galaxies are in space with lensing surveys to show us how their images are distorted on the sky, we are mapping out where the stuff is in the universe. This is something that literally was first done in

2007. It's a technique that we're going to be perfecting and improving in the next years to come.

The great leap forward in doing these kinds of surveys will be something called the LSST, the Large Synoptic Survey Telescope. This is a telescope that has been on the drawing board for quite a while now and will be for quite a while to come. It is supposed to start construction in 2010, be finished by 2013, being built in the mountains in Chile where the air is very thin, so you get a very good view of the sky. The goal of the LSST will be to survey a large patch of the sky every single night. It's something like the Sloan Survey does, is over the course of years it takes images of different patches of the sky and pieces them together into an image of what is going on. The LSST will basically be doing that every day, so it will be collecting a huge amount of data, unprecedented in astrophysics— over a pedabyte of data per year. Just like a gigabyte is 1,000 megabytes, a terabyte is 1,000 gigabytes, and a pedabyte is 1,000 terabytes. One pedabyte is the equivalent of 250 billion songs on your iPod, a huge amount of data, so much so that just surfing through the data looking for different pieces of information is an incredibly difficult computer science problem.

In fact, you will not be surprised to learn that scientists working on the LSST project are collaborating with researchers from Google, who are the best people in the world at searching through large databases and finding things, so astronomers want Google's technology to search through their database. Google wants the data that astronomers have because astronomers are the only people in the world who collect huge amounts of data and then give it away for free.

What we're going to be getting from LSST is huge amounts of data. Lensing will be an important goal, to be able to look for the 3D structures of the universe, but the other thing is that we'll be looking at the same region of the sky over and over again. In other words, we will be collecting data in the time domain, something that astrophysicists have traditionally not done because it was just too expensive. In other words, we will not just get weak lensing and large-scale structure, but we will also discover supernovae. Furthermore, we will discover asteroids that are coming to crash into the earth. It is a major funding source for the LSST that will find objects in the solar system that might be a danger to us. It doesn't tell us anything about dark matter and dark energy, but it's still a useful goal.

Going to smaller and smaller things in the universe, next let's consider clusters of galaxies. Remember, clusters are, on the one hand, the largest bound structures in the universe. On the other hand, they are a fair sample of what is in the universe. By studying clusters so far, we have been able to learn about the relative proportions of, let's say, dark matter to ordinary matter in the universe.

The new thing we'll start doing with clusters of galaxies is to count them. By counting the number density of stuff in the universe, you're tracing out how the universe has evolved. People have always known this is true, but historically, this is a very difficult technique to use because when you're counting things, it's crucially important you don't miss anything. In astrophysics, some things are bright and some things are dim. When you try to count things, you don't know that you've found all of them. But, clusters of galaxies are special. In a cluster, the galaxies fall together, gas and dust falls in along with the galaxies, and the gas and dust heats up. There is something called the *Sunyaev-Zeldovich Effect* in which it's not that we're looking at the gas, but we're looking at the shadow cast by this hot gas on the cosmic microwave background. You look at the microwave background in the sky. You look for a little dot on it, a little dark spot. That dark spot is caused by the fact that the photons from the microwave background come into the cluster, scatter off the gas, and don't get to you. The reason why this is important is because if you were just looking at the light emitted by the cluster, it would become dimmer and dimmer, as you get further and further away. So, it's harder to find faraway clusters. But, if you're looking at the shadow that is cast by the cluster, it's a shadow. It doesn't get any dimmer the further away you get. By doing a survey for the Sunyaev-Zeldovich Effect, you will be able to find all of the clusters of galaxies in the patch of sky on which you're looking, an unbiased sample of all the clusters of galaxies in some region of the universe.

We are currently doing that with something called the South Pole Telescope, a 10-meter radial telescope set up in Antarctica at the South Pole that starts operation in 2007. Once the South Pole Telescope finds clusters of galaxies, we will then follow up with X-ray observations. It's X-rays that will allow us to actually weigh these clusters of galaxies and measure how much mass is in them.

More importantly, it's the number of clusters of galaxies that count. The dark energy, making the universe expand as the dark energy takes over, it stops structures from forming. When you have a matter-dominated universe, the matter gradually collects together and forms bound structures. But, when the dark energy takes over, the matter is pushed apart and galaxies stop falling together. So, the number of clusters of galaxies as a function of distance in the universe is a very sharp way of figuring out how much dark energy there is. It is arguably competitive with supernovae and other measurements. We don't know yet because we still don't understand the fundamental physics of clusters of galaxies very well, but that's something that numerical simulations are teaching us about right now.

Meanwhile, of course, we're not done with supernovae. Supernovae are the way we discovered the dark energy in the first place. We want to do better. The important role there is that we would like to understand the fundamental intrinsic nature of the supernovae themselves. If you collect enough examples—instead of 20, or 30, or 40 supernovae, if you have 1,000 supernovae—then you can start talking about what the distance redshift relationship is for each individual possible kind of supernovae, what it looks like for supernovae in elliptical galaxies, what it looks like in spiral galaxies, and so forth. You can search for possible systematic effects that might put you on the wrong track. By doing this, we'll be able to measure the equation-of-state parameter of the dark energy—what we call w, what should be -1 if the dark energy is truly constant, and something like -.8 or -.9 if it's some gradually declining quintessence. So, looking for supernovae at very high redshifts is our best hope for figuring out whether the dark energy is truly constant or slowly varying.

Finding supernovae at large distances is hard if you're here on earth; therefore, there are proposals to build a dedicated satellite to do exactly that. Right now, the plans are for NASA and the Department of Energy to team up to build something called the Joint Dark Energy Mission (JDEM); however, I can't promise you it will be built. These things depend on things like Congress and money, and other things that scientists don't have much control over.

Finally, in terms of looking at photons, we can look at gamma rays to look for evidence for indirect dark matter. That is to say, we have dark matter particles in the universe; we can of course look for them here on earth, but we can also look for them in the sky by waiting for dark matter particles and dark matter anti-particles to annihilate. If the dark matter is a weakly interacting massive particle, then the real dark matter density comes from both particles and anti-particles. Occasionally, they are going to annihilate and give off gamma rays. Right now, we don't have much of a capability for looking at gamma rays in the sky, but the new GLAST satellite is scheduled to be launched in late 2007. It will give us our first direct view and might lead to an indirect detection of dark matter.

Of course, those are not the only particles we use. We don't only use photons to look at the universe these days. We use other kinds of particles. For example, we use neutrinos. Ever since the solar neutrino problem was noticed in the late '60s–early '70s, we've been able to find neutrinos from outer space, not just neutrinos here on earth. Besides the sun, the one kind of neutrino that we've been able to detect from outer space is neutrinos from supernova 1987a. One of the things that neutrino physicists have been doing is planning for the next supernova. They are ready to get even more information about fundamental neutrino properties from whatever supernova goes off next within our own galaxy. We don't know when that's going to be, but it typically happens once every 100 years, so you never know.

Meanwhile, we'd like to learn more about the fundamental nature of neutrinos, which has recently opened up as a growth field within physics. We now know that neutrinos have mass. They are very light, but they have some mass. Because they have mass, they can mix. In other words, we said that there were things called electron neutrinos, things called muon neutrinos, things called tau neutrinos, but these neutrinos can change into one another as long as they have mass. This has been experimentally found and it's the kind of thing that we're trying to understand better and better. The role for dark matter is that none of the three neutrinos we know and love can be the dark matter, but there could be others. There could be kinds of neutrinos that we haven't yet directly detected—for example, sterile neutrinos, which do not interact with W and Z bosons. The only way that sterile neutrinos interact is through gravity and with other neutrinos. We're trying to build neutrino detectors that will be able to tell the difference between sterile neutrinos and ordinary neutrinos because that might play an important role in the dark matter mystery.

Getting more and more exotic, we can look at gravitational waves. Just like electromagnetic waves arise from electric charges moving up and down, gravitational waves are created when heavy objects move back and forth. For example, if you have two objects orbiting each other they will be emitting gravitational waves. Gravity is a very weak force. These waves are not very easy to notice. We have never yet directly detected gravitational radiation. We do have an observatory that is trying to do exactly that. We have LIGO (the Laser Interferometer Gravitational-Wave Observatory), which finally right now in 2007 has begun to collect data at the rate it was designed to do. It has not yet detected any objects, and it may or may not succeed in doing so.

We have a better plan for a gravitational wave observatory in space. This is called LISA (the Laser Interferometric Space Antenna), which will be three satellites, five million kilometers apart from each other. They will be looking for gravitational waves in a completely different wavelength than LIGO is looking, and that's actually good news for cosmology. One of the things that LISA will be looking for is that you have a tiny black hole orbiting a supermassive black hole—a million solar mass black hole—and it will be orbiting, in-spiraling, giving off gravitational waves all along. It turns out that just like supernovae or standard candles, in-spiraling black holes can be standard sirens. You can learn enough about the parameters of the black hole binary to figure out how much gravitational waves were given off at the source. By detecting them here and measuring how strong the gravitational waves appear to us, we can figure out how far away it is. This is a completely independent way to measure distances in cosmology, and a very, very clean way, one that doesn't get messed up with the messy astrophysics that ordinary objects giving off photons do. If we can turn standard sirens into a useful cosmological tool, we'll be able to get unprecedented precision in mapping out the expansion of the universe out to very distant redshifts.

Next up, we have dark matter. Remember, we can look for dark matter by using photons, by looking at the gamma rays that are emitted when dark matter and anti-dark matter particles come together. But, of course, we can also look for dark matter right here on earth. We can build underground detectors that will wait for the dark matter particle to come in, bounce off the nucleus in the detector, and leave a little bit of energy there. We have detectors working right now. Unfortunately, we don't have a very precise prediction for when they will be able to observe things. There is a lot of freedom in the predictions there, so we can't say what year by which we should be able to detect things, but it could very well be any moment. There is a detector called CDMS (Cryogenic Dark Matter Search), which is underground in Minnesota and working right now and currently has the world's best limits.

There is another detector called DAMA in Italy, which works on an interesting system where it doesn't look for the total amount of signal. It looks for the variation in signal throughout the year, as the earth revolves around the sun and moves through the dark matter cloud around it. The reason why I mention DAMA is because DAMA has claimed to find a signal. They claim that they have already detected the dark matter. The reason why this is not earth-shattering news is because most people don't believe them. The current consensus in the dark matter detection community is that they have some systematic error they haven't yet detected. But, there is an interesting alternative possibility. It could be that there is a new kind of dark matter, some specific particle physics model, which shows up more easily in something like DAMA than something like CDMS and the other detectors. In other words, this is why we do experiments. We do experiments because the real world might be more complicated than the simple models that we write down. So, stay tuned for news from dark matter detection experiments. You might be very quickly learning important features about 25 percent of the universe.

Coming in closer to home, besides looking at things in the sky, we can also do experiments closer by. Before, ordinarily, closer by would mean here on earth in the laboratory, but these days you have to extend it to at least the solar system. We can send satellites throughout the solar system to help us do physics experiments. The most obvious example is testing general relativity. We already mentioned the Cassini Mission, which primarily had as its goal to take pictures of Saturn and Jupiter and the other outer planets, but also tested Einstein's theory of general relativity by sending signals to us that we measured the time delay of. There are new upcoming missions that will also test general relativity. There is something called the STEP Mission. The Satellite Test of the Equivalence Principle will be asking if different objects made of different stuff move differently in a gravitational field.

Remember, a very firm prediction of Einstein's theory of general relativity is that everything responds to gravity in exactly the same way. That is known as the *equivalence principle*. If there is a fifth force, if there is some new light field that stretches out over macroscopic distances, it will show up as a violation of the equivalence principle. Quintessence might be exactly such a field. We don't know; but, if quintessence exists, it's a very light boson that could give rise to forces that violate the equivalence principle. Experiments like STEP will be looking for exactly this. It's an example of a high-risk, high-payoff thing. Probably, they will not see anything. But, if they see something, it would truly change the way we think about the forces of nature.

Of course, we're doing similar things here on earth. We're looking for new forces and for new features of old forces. We're doing experiments here on earth to look for violations of the equivalence principle, looking for different objects being accelerated by different amounts in the gravitational fields of the sun or the earth. We're also looking for violations of the inverse-square law of gravitational dynamics. When we talked about string theory and extra dimensions, we mentioned the possibility that some of these extra dimensions could be pretty big. They could be as big as a millimeter across. Right now, we've been able to squeeze down the limit on the size of the extra dimensions down to one-tenth of a millimeter. We're going to continue to try to squeeze that down more and more. This is, again, an example of an experiment for which if you find something, you absolutely win the Nobel Prize. You've found something incredibly important in the history of physics. The smart money bets they won't find anything, but you don't know until you look. We need to have some experiments that spend their time on the high-risk projects.

Finally, let me emphasize the role of particle physics experiments in the future of cosmology. We've been talking about dark matter and dark energy—95 percent of the universe. The evidence that we have for dark matter and dark energy comes from looking at the sky, from doing astrophysics and taking observations of galaxies and clusters in the microwave background, and supernovae, and the whole universe all at once. But, ultimately, the explanation for dark matter and dark energy will have to come from particle physics. The dark matter is some kind of particle, either an axion or a supersymmetric particle, or something. The dark energy is even more mysterious, either quintessence or vacuum energy or something even worse. To get a good idea of which of those models is more plausible than the others, we need to understand particle physics beyond the Standard Model. So, doing laboratory experiments here on earth to increase our understanding of particle physics is probably the single most promising way to learn more about the dark sector.

Right now, we have the Tevatron at Fermilab just outside Chicago as the highest energy particle accelerator on earth, but later in 2007, the LHC, the *Large Hadron Collider* at CERN, outside Geneva, will turn on. Starting in 2008, CERN and the LHC will be doing high-energy particle physics in a new realm where we have never looked before. We're going to be looking at exactly the regime where we should be able to find the Higgs boson, supersymmetric particles, and other particles just beyond the Standard Model. Not only will we be looking for things that might very well be dark matter, and might very well therefore be 25 percent of the universe, we'll be hoping to learn things about supersymmetry and extra dimensions. We'll be hoping to learn things about phenomena that come into the discussion when we talk about dark energy.

I realize that's a very vague construction. I didn't promise you anything very specific right there, but the point is that dark energy is so mysterious we need all the help we can get. There is no known way that we can think of to do a particle physics experiment to detect dark energy or to see exactly what it is. But, by learning more about the big picture, something that seems very unnatural to us might suddenly make sense, so we're extremely hopeful that particle physics over the next five years will be surprising us. We'll be teaching new things about the fundamental nature of space and time and matter, which ultimately will illuminate the nature of 95 percent of the universe we live in.

Lecture Twenty-Four
The Past and Future of the Dark Side

Scope:

The concordance cosmology is an excellent fit to a variety of data but presents us with deep puzzles: What are the dark matter and dark energy? Why do they have the densities they do? Our own universe seems unnatural to us. That's good news, because it is a clue to the next level of understanding. Scientists are never content; they are always searching for a more accurate and comprehensive model of nature. Things that today seem puzzling—dark matter and dark energy foremost among them—will tomorrow serve as crucial ingredients for a better picture of the universe.

Outline

I. Let's recap how we got where we are—a quick historical overview reveals how far we've come in understanding the universe.

    A. A century ago, we didn't know how stars burned, that there were other galaxies, or that the universe was expanding. We certainly didn't know anything about dark matter and dark energy.

    B. Since then, we've put together a remarkably successful picture of the universe on large scales, as well as a greatly improved understanding of fundamental physics on small scales

        1. A century ago, we didn't know how stars burned, that there were other galaxies, or that the universe was expanding. We certainly didn't know anything about dark matter and dark energy.

        2. Since then, we've put together a remarkably successful picture of the universe on large scales, as well as a greatly improved understanding of fundamental physics on small scales.

II. Our current, extremely successful, model of the universe has three basic features.

A. The universe is expanding, and understanding this expansion within the context of general relativity, we can extrapolate all the way back to within one minute of the Big Bang.

    1. We live in a smooth, expanding universe that originated in a hot, dense state.

    2. Predictions of the Big Bang model, including nucleosynthesis and cosmic microwave background anisotropies, have confirmed the general picture beyond a reasonable doubt.

B. An inventory of the universe has been pieced together in recent years: 5 percent ordinary matter, 25 percent dark matter, and 70 percent dark energy.

    1. Dark matter is cold and (thus far) noninteracting; dark energy is very smoothly distributed and remains nearly constant in density as the universe expands.

    2. Again, a wide variety of observations are consistent with this inventory.

C. Finally, we have the story of the growth of structure, from small perturbations at early times into the stars, galaxies, and clusters that we observe today. This picture does not absolutely require a mechanism for generating the perturbations in the first place, but we have one: the inflationary universe scenario.

III. Why is the universe like it is?

A. Reflecting upon whether or not the current picture of the universe makes sense to us at a deep level—whether it is somehow natural to us—is another way (beyond the experiments discussed in Lecture Twenty-Three) to think about how we can go beyond our current understanding.

    1. A configuration of matter is natural if it's robust; for example, if you move some particles around, you will nonetheless get the same basic outcome.

    2. This kind of reasoning is not just philosophizing. It can lead us to very explicit scenarios that help us explain something that we don't otherwise understand.

    3. For dark matter, we have scenarios that make dark matter seem very natural.

4. For dark energy, however, we do not have any models that would pass the naturalness test.

   a. The simplest model for dark energy, vacuum energy, fits what we see going on only at the cost of an extremely unnatural parameter.

   b. The vacuum energy model also does not explain the puzzling coincidence that the amounts of matter and of dark energy are today roughly the same, even though back when the microwave background was formed, the density of dark energy was only one-one billionth of the density of matter.

B. One possible route toward future progress lies in the physics of the dark sector. As far as we can tell at present, dark matter and dark energy interact only through gravity, but the possibility of much richer physics lurks beneath.

   1. The dark energy could interact with itself: Rather than being absolutely constant everywhere, it might change slowly with time, it might change slowly with space, and it might have perturbations in its density.

   2. The dark matter might interact with itself—in fact, every realistic model of dark matter includes some interactions between the dark matter and itself.

   3. Ordinary matter and the dark sector might interact with one another.

      a. The dark matter might feel the weak force, or some other force, of ordinary matter

      b. If the dark energy is vacuum energy, then it will not interact, but if the dark energy is dynamical, then we might find interactions between the dark energy and photons, or between the dark energy and nuclear matter.

   4. The dark matter and dark energy might interact with one another independently of ordinary matter.

   5. These issues are begging to be addressed experimentally, and a broad-based program is under development to do just that.

IV. It seems as if we know a lot, but we have a lot of unanswered questions. That, of course, is the perfect situation to be in for a scientist! It means that we're not simply flailing in the dark, but much remains to be discovered.

A. The apparently unnatural state of our current universe provokes a dramatic speculation.

   1. The most obvious feature of our current universe—so pervasive that we haven't even mentioned it yet—is the difference between the past and the future.

   2. Physicists quantify this difference in terms of *entropy*—a measure of disorder that grows inexorably with time. Why was the entropy so small in the early universe, near the Big Bang?

   3. Perhaps the answer has something to do with dark energy. Vacuum energy has the property that it keeps space from being completely quiescent; there is always a tiny temperature.

   4. If we wait long enough (much longer than the current age of the universe), fluctuations in an otherwise empty space could create a small pocket of dark energy, similar to what is needed to begin inflation.

   5. It's conceivable that such an event, in a preexisting universe before our Big Bang, is the origin of everything we see.

   6. Such speculation may play a role in helping us understand why the configuration of our current universe looks as though it started in such an unnaturally low state of entropy—this is the kind of speculation we're led to as a result of thinking about dark matter and dark energy.

B. A century ago, some scientists were convinced that a final codification of the laws of nature was just around the corner.

   1. Not everything fit together, but most things did, and the few remaining puzzles were simply details to be ironed out. Those details, of course, grew from minor irritations to major features of our understanding of the universe, as the twin revolutions of relativity and quantum mechanics transformed the way we think about physics.

   2. It could be that a similar revolution or two is just around the corner, presaged by the puzzles of dark matter and dark energy. Or perhaps not—there may be quite prosaic explanations for us to settle on. Only further experiments will tell us, and the uncertainly is half the fun.

Recommended Reading:

Greene, *The Fabric of the Cosmos*, chapters 10–11.

Livio, *The Accelerating Universe*, chapters 7–10.

Vilenkin, *Many Worlds in One*, chapters 16–19.

Questions to Consider:

1. Science doesn't "prove" theories rigorously, as mathematics does. How should we judge when a theory is "right enough" to be accepted, as opposed to simply being a tentative hypothesis?

2. What would count, in your mind, as an explanation for why our universe is the way it is?

3. We've learned a lot about cosmology in the last 100 years. What do you think we might learn over the next 100 years?

Lecture Twenty-Four—Transcript
The Past and Future of the Dark Side

At this point in the final lecture of the course, we have a great deal to be proud of. By "we," I mean scientists who have been trying their best to put together a picture of what the universe is made of over the last 100 years. But, I also mean the human race that has been wondering about these questions for the last several thousand years, and of course all of you who have been watching these lectures.

We've put together a pretty good picture of everything. Not only do we know that 95 percent of the universe is this dark sector—70 percent of it dark energy that is smoothly distributed through space and time, and 25 percent of it dark matter, some heavy particle that settles into galaxies and clusters—but we have good reason to believe that we are not missing anything. There may be one percent or more here or there, but basically we know everything in the universe. Measuring the total density of stuff through the curvature of space convinces us that what we know right now comprises everything that there is to know.

However, scientists are notoriously bad at resting on their laurels. What we want to do is to focus on the questions to which we don't know the answers, the puzzles that we're left with by this very, very successful picture. Fortunately, even though the picture that we have of the universe today is very successful, it leaves us with tremendous questions and it offers us clues as to where we might find the answers. The clues reside in the fact that this picture that fits the data so well is nevertheless extremely unnatural from certain points of view.

It's perfectly fair to say that our picture of the universe—this 5 percent ordinary matter, 25 percent dark matter, and 70 percent dark energy—makes no sense, or at the very least you could have done better if you were asked to design a simple and elegant universe. So, there is something that we don't understand about why our universe is like this versus like something else.

For this lecture, what we're going to do is to step back a bit, and look at the big picture. Start with a historical overview of how we got here, and try to go from there to extrapolate where we might go next to try to understand—given what we have learned about the universe—where might we be looking for clues as to what underlies the particular configuration in which we find the universe in which we live.

Let's start with how we got here. A hundred years ago we knew next to nothing correct about what the universe was like on very, very large scales. We did know that there were stars. We knew that stars were different than the sun. They were further away, but they were the same basic kind of thing. We didn't know what those stars were. We didn't know how they made their nuclear fuel—the stars or the sun. We didn't know that there was such a thing as a nucleus. We knew there were atoms. We knew there were different chemical elements, but we hadn't yet decomposed atoms into electrons moving around nuclei. So, the basic energy source for these stars was unknown to us. We believed that the stars were distributed in an "island universe" configuration. We thought that the stars were moving around each other in what we now call the Milky Way Galaxy, but we didn't know that there were other galaxies. We knew that there were little fuzzy patches on the sky that were called nebulae. We didn't know that many of these nebulae were in fact galaxies in their own right, with hundreds of billions of stars all by themselves.

We certainly didn't know about the expansion of the universe. We didn't know that galaxies had an apparent velocity moving away from us. We didn't know anything about Einstein's theory of general relativity, so we didn't know that space and time could expand. We believed in an absolute Newtonian universe in which space and time were fixed. They weren't changing as a function of time, so the dynamic nature of spacetime itself was something that we didn't have access to.

The modern story really begins in 1915 when Einstein put the finishing touches on the general theory of relativity, which says that spacetime can be curved. It can have a dynamic all of its own, and that dynamic is manifested to us as gravity. He very soon thereafter started thinking about cosmology and realized that if the universe were evenly spread all over the place, it would have to be dynamical. It would have to either be expanding or contracting, but as far as he knew in 1917, it wasn't. So, he changed his ideas. He added a new term to his equation for general relativity called the cosmological constant, something that would keep the universe static, at least in principle.

Then, of course, in the late 1920s, it was realized the universe is not static. Hubble taught us, number one, that some of the nebulae are separate galaxies all by themselves and, number two, that the universe is expanding—that is getting bigger. Einstein tossed away his cosmological constant, realizing that it's a dynamical universe in which we live. Very soon thereafter, Fritz Zwicky was looking at clusters of galaxies and he found the first evidence for dark matter. He looked in the Coma Cluster and realized there was more stuff in the Coma Cluster than we could possibly account for by what we saw. He didn't know at the time that that stuff, that dark matter, couldn't be explained by ordinary atoms, but that was going to come down the road.

In the 1940s and '50s was when we first really realized how the early universe worked. George Gamow and his students, Ralph Alpher and Robert Herman, figured out how to make light elements by nucleosynthesis in the early universe. As a spin-off, they got the cosmic microwave background—the leftover radiation that we now know provides us with a snapshot of what the universe looked like 400,000 years after the Big Bang. This microwave background was discovered in 1960 by Penzias and Wilson by accident. They were very careful radio astronomers. They kept getting these microwaves uniformly across the sky. They didn't realize there was another group down the street in Princeton that was looking for these intentionally. They won the Nobel Prize in 1970 for their work.

Then, in the '70s, particle physicists put together the Standard Model of Particle Physics. So, the cosmological Standard Model crept up on us gradually. The Particle Physics Standard Model was put together over the space of only a few years and it's been extremely successful since then. Since the mid-1970s, no particle physics experiment has truly surprised us. Everything that we've done has fit together into what we already knew as the Standard Model.

Also, in the 1970s, Vera Rubin found even better evidence for dark matter. She measured the mass of individual galaxies, looked at their rotation curves, and found that not only are the galaxies heavier than they could be given only the stars and gas and dust that were in them, but that there was mass out beyond any of the visible matter. Eventually, we realized that this dark matter was not able to be accommodated in what we understood about particle physics. It had to be a new particle.

In 1980, Alan Guth invented inflation. There were puzzles about the early universe we didn't understand. How is the early universe so smooth? How is the spatial geometry of the universe so close to flat? Inflation made all that snap into place and even though we are still not absolutely sure it's true, it's still the leading candidate to explain why the overall geometry of our universe is what it seems to be.

Then, in the 1990s, observational cosmology became a precision science. The one watershed moment was the discovery in 1992 of the tiny fluctuations in temperature in the cosmic microwave background. This was the result from the COBE satellite of NASA. What this did was enable us to begin linking the earliest moments of the universe's history to what we observe today. At early times, the universe was extremely smooth. One part in 100,000 was only the kind of deviations that you could see. Under the force of gravity, those deviations have grown into galaxies and clusters that we observe today. By now, modern technology allows us to observe those galaxies and clusters to exquisite precision, to compare what we see to the predictions that you would get from the cosmic microwave background extrapolated in time, and with that to put together the ingredients of the cosmological model that is so successful.

The final ingredient that we need was found in 1998. Supernovae used as standard candles to discover that the universe is accelerating, which can only be explained by dark energy, by some form of stuff whose energy density doesn't go away as the universe expands. That made up the extra 70 percent that was predicted by inflation, if you wanted to live in a flat universe.

Then, in the year 2000, observations of the cosmic microwave background confirmed that indeed we do live in a flat universe. The fact that 70 percent of the universe is dark energy and 30 percent of it is dark matter simultaneously fits the dynamical measurements of galaxies in clusters, the supernova data that tells us the universe is accelerating, and the cosmic microwave background that tells us that the universe is spatially flat.

Since the year 2000, we've been confirming this picture in ever more elaborate ways. The WMAP satellite has given us extra information that in fact the ordinary matter is only five percent, that we need some form of dark matter. The evolution of structure from then to now is consistent with what we would think it would be in this model with a universe of 70 percent dark energy.

We land here at a place where we have a very successful model of everything in the universe. If you think about it, this model has three basic features—not just the three ingredients in the inventory, but three aspects that are important to its success. Number one, the fact that the universe is expanding and the understanding of this expansion within the context of general relativity. In the 1915–17 era, Einstein was realizing that the universe had to be expanding. We now know that it's true. We can use his equation to extrapolate all the way back to within a minute of the Big Bang, and we get the right answer as it's confirmed by Big Bang nucleosynthesis. The second feature is, of course, the inventory—the 5-25-70 split between ordinary matter, dark matter, and dark energy, which has been tested in multiple ways and is over-determined by the observations we have now. The final aspect is the evolution of structure of the universe. The first two aspects deal with a universe that is spatially smooth, more or less the same everywhere. It is an extremely accurate picture, either at early times or on large scales today. But, on smaller scales, we see that there are galaxies, there are clusters, there are stars and planets, and fortunately we are able to understand where they came from going from the early universe to now.

That's an incredibly successful picture. We can go using data from one second after the Big Bang to 14 billion years after the Big Bang. We can extrapolate with our theories sensibly back to times like 10^{-30} seconds after the Big Bang when we talk about inflation, and inflation acts as a way to generate the perturbations that we see today. When we are hypothesizing and speculating about what the universe was like 10^{-30} seconds after the Big Bang, we can actually connect those speculations to observations we are making today. We're not absolutely sure of what the universe was like earlier than one second old, but the very fact that we're trying to go back from one second to some tiny fraction of a second is indicative of the progress we've made compared to 100 years ago when we didn't know that the universe was even getting any bigger.

Let's step back and ask, given this extremely successful picture, why is it like that? Why is the universe smooth and homogeneous? Why is it that 70 percent of the universe is dark energy versus the 30 percent that is matter? You can ask yourself, is this a respectable question to be pondering? Should we be people who just look at the universe, describe what it's doing, and that is our duty as scientists? It's not our job to ask why it is like that. But, of course, as scientists, we are not satisfied with the current state of affairs. We always want a more complete picture. One of the ways to help us get a more complete picture is to make more experiments, to collect more data, to take more observations, and to find out more about the universe. We're certainly going to be doing that. In a previous lecture, we gave you this very diverse smorgasbord of experiments that will be done, a portfolio of different things, both highly speculative and guaranteed to get good results.

In the meantime, there is another way to think about how we can go beyond our current understanding. That's to try to reflect upon whether or not the current picture of the universe makes sense to us at a deep level, whether it is somehow natural to us. There are no absolutely strong contradictions in between the different elements of the Standard Model of cosmology, like Einstein who was lucky enough to have a deep contradiction between Newtonian gravity and special relativity when he was developing general relativity. Yet, there are tensions between the different elements. We're going to try to ask is it somehow a natural universe in which we live, and if not, is the fact that the universe appears unnatural to us a sign that there are some deeper underlying forces with respect to which it would seem much more natural? We, as scientists, have ways to think about this concept of naturalness, ways to take this more or less fuzzy and human idea and turn it into something a little bit more quantitative. Both for theories that we have and for situations we find in the world, scientists can tell you whether or not that particular thing seems natural to us.

For a theory like the Standard Model of Particle Physics, for example, we say it's natural if everything that might be happening in that theory is happening, and it's happening with about the right strength. That didn't seem much less fuzzy than the original idea of naturalness. What we mean quantitatively is that when we look at the numbers that appear in that theory, the numbers that characterize the strength of different forces and the frequency of different interactions, those numbers don't seem very finely tuned. In other words, if we were to change those numbers just a little bit, you would get the same basic kind of physics within your theory.

But, when we look at some different aspects of the Standard Model, we see that there are fine tunings there. One example is what particle physicists call the hierarchy problem. They look at the mass of the Higgs boson. We haven't detected it yet, but it's probably something like 100–150 times the mass of the proton. But, if you look at it, there is no reason why that number, the mass of the Higgs boson, shouldn't be much, much larger, shouldn't be up there near the Planck scale—10^{18} times the mass of the proton. Why is the mass of the Higgs boson so small compared to the natural field that it would likely have? That's an example of the kind of naturalness argument that particle physicists use. What they will then do is try to find some underlying reason; in fact, in the form of supersymmetry, there is an underlying dynamical explanation for why the Higgs boson might be so much lighter than the much heavier scales characteristic of the rest of particle physics.

For situations in the universe, there is a slightly different aspect of naturalness with which we're concerned. A configuration of stuff in the universe is natural if it's robust. If you move some particles around, you will get the same basic idea. This is the kind of thing you might have been exposed to in some physics class a long time ago. If you have a box full of gas and if all of that gas happens to be stuck in one corner of the box, that's not a very natural configuration. You let it go and it begins to fill the box. You say that all of the air molecules being stuck in one corner have a very low entropy, and as you let it go, it's going to naturally fill the box. If you start with the gas filling the box, it will not naturally curl up in one corner. The configuration in which stuff is all spread out is just much more robust. You can move a few particles around. It's still gas smoothly spread out throughout the box.

So, we have these notions. We have these notions of naturalness and unnaturalness. It is not that if you observe a gas with all the particles in one corner of a box, you would say that violates the laws of physics. It's completely consistent with the laws of physics, but it's unnatural. We say there must be some reason why it's there. There must be some dynamical physics mechanism why it looks like that. We want to apply that same kind of reasoning to the whole universe. This is a successful strategy that has been pursued before. For example, the horizon and flatness problems were inspirations for the development of the model of inflation. The horizon and flatness problems were, number one, the idea that the universe is smooth on a scale that's larger than it has any right to be. If you look at the universe in the microwave background era, there are points with almost exactly the same density and temperature, but they shared no common points in their past. There was no way for them to know to be the same temperature and yet they are. That's a naturalness problem.

Space, meanwhile, is very, very close to geometrically flat, but tends to become less and less flat as time goes on. Why was it so close to flat in the early universe? That's another naturalness problem. By thinking about these problems, scientists were able to invent the model of inflation. Inflation solves these problems and, as a bonus, provides an explanation for the perturbation that we now think grow into large-scale structure. So, this kind of reasoning is not just philosophizing. It can lead us to very explicit scenarios that help us explain something that we don't otherwise understand.

What can we do? What if our current universe doesn't seem natural to us? Do we have possible mechanisms that make it fit into a bigger picture? The two things we're concerned about in this set of lectures are dark matter and dark energy. The dark matter and dark energy is a situation that is actually very different in the two cases. For the case of dark matter, we have scenarios that make dark matter seem very natural to us. We have, first and foremost, the WIMP scenario, the Weakly Interacting Massive Particles. If there exists in nature some particle that is heavy, that is electrically neutral, so that you can't observe it very easily, but that feels the weak nuclear force, as long as that particle is stable, it will naturally give rise to a density of particles that is very similar to what we observe in dark matter today. In other words, you don't have to work very hard to extend the Standard Model of Particle Physics so that naturally a dark matter candidate appears with naturally the right amount of stuff in the universe.

We don't necessarily know that that's the right scenario, but it's a scenario that fits in with our understanding of what probably would make sense to us. Therefore, we take it seriously, we take this idea that the dark matter is a Weakly Interacting Massive Particle seriously enough that we build very expensive experiments to go look for it on the basis of that hypothesis. We build particle accelerators to try to produce it directly. We build satellites to try to look for the *annihilation* of WIMPs and anti-WIMPs in space, and we build underground detectors to find them directly in our own laboratories.

Meanwhile, for dark energy, we do not have any models that would pass the naturalness test. The simplest model for dark energy, since we know it is smoothly distributed through space and nearly constant as a function of time, is vacuum energy. Vacuum energy is something that is absolutely the same, the same density of energy at every cubic centimeter of space and time, an absolutely minimum energy that doesn't devolve or fluctuate in any way through time or space. It's a simple model, but it's not natural in terms of what we know about dark energy because it doesn't have the right value. You can estimate how big the vacuum energy should be. According to our current understanding of gravitation and of particle physics, the vacuum energy should be enormously bigger than what we observe. It should be 10^{120} times what you actually get. So, we don't know what is going on. We have a theory that fits, but only at the cost of an extremely unnatural parameter.

The other thing that we know about the vacuum energy is that is victim to something called the coincidence scandal. If you'll notice in the current universe, the dark energy is about 70 percent of the total energy budget. Matter is about 30 percent. Now, by cosmology standards, 70 percent and 30 percent are equal. Those are very, very similar numbers. The reason why this is a little bit of a surprise is because dark energy and matter evolve very differently with respect to each other. Matter dilutes away as the universe expands, so that density in matter when the microwave background was formed was a billion times bigger than the density now. In the early universe, when things squeezed together, the densities were much higher.

Dark energy, meanwhile, has a density that is constant. The number of ergs per cubic centimeter is a tiny number that is fixed. It was just as small in the early universe as it is today, as it will be in the future. Back when the microwave background was formed, the density of dark energy was the same as it is today. In other words, it was one-one billionth of the density of matter. These two numbers—the energy density in dark energy and the energy density in matter—change rapidly with respect to each other as the universe expands.

Yet, today, when we happen to be here looking for them, they are approximately equal. Why is that? We don't know the answer to this question. It might be explained by environmental selection. It might be that most observers in some large ensemble of the multiverse are going to be born at a time in their regions of space history when the density of dark energy is similar to the density of dark matter. That is a hypothesis. That's the kind of thing that we're driven to contemplate by thinking about these naturalness arguments. It may or may not be right, but these give us a clue for the directions in which to move next.

Here is a picture that we can return to of where we might go in future years after we've studied the dark sector in greater detail. We right now have a very successful understanding of ordinary matter, of the Standard Model of Particle Physics. We have dark matter and dark energy, which are consistent with the data if they only interact with us through gravity. But, that doesn't mean that's the final word. It could be much more rich than that. So, there are plenty of possibilities for new physics here that we're looking for in future experiments.

The dark energy could interact with itself. In other words, it might not be absolutely constant everywhere. It might evolve. It might change slowly with time in one way or another. It might also change slowly with space. There could be perturbations in the density of dark energy. Likewise, the dark matter might interact with itself. In fact, in every realistic model of dark matter, there are some interactions between the dark matter and itself. There is some way for the dark matter to be produced in the first place. More interestingly, there could be interactions between ordinary matter and the dark sector, one way or another. There could be interactions between dark matter and ordinary matter like the kinds that we're looking for. It could very well be that the dark matter feels the weak interaction of the Standard Model of Particle Physics. That would enable us to create the dark matter in accelerator experiments and in deep underground laboratories. Even if not, even if the dark matter is not weakly interacting, it might interact with some other force.

The axion is an example of a particle that is electrically neutral, yet indirectly interacts with photons. We have separate experiments that are dedicated to looking for the interactions of axions with photons. It could be any day now people will begin to actually discover the dark matter directly because of those interactions. We are looking for them because they make the dark matter theories more natural in the context of other things that we understand about particle physics.

Best of all, maybe, would be if the dark energy directly interacted with ordinary matter. If the dark energy is just vacuum energy, if it's just a constant amount of stuff in every cubic centimeter, then it will not interact. It's just a background feature of spacetime and all we'll ever be able to do is measure its effect on the curvature of spacetime, on the expansion of the universe. But, if the dark energy is something different, if the dark energy is dynamical, something that changes from place to place, then dynamical fields tend to interact. We can look for interactions between the dark energy and photons, or the dark energy in nuclear matter, and experiments are ongoing to do exactly that.

It could even be that the dark energy has similar kinds of interactions with the dark matter. It could be that the dark sector all by itself forms some interestingly interacting set of things and we, the ordinary matter, are sitting off by the side. The dark matter could be affected by the dark energy and have its mass change, or it could have its interactions changed in some interesting way. We don't know, but it's not bad that we don't know. It's good that there are so many possibilities that we're going to be pursuing, looking for observational evidence for any one of these things happening.

For the last part of this set of lectures, I want to end with one dramatic speculation and one cautionary tale just to bring us back to reality at the end. The dramatic speculation has to do with, once again, the naturalness of our picture of cosmology. In this case, I'm not talking about the naturalness of the theory that we have—the theory with the parameters that describe dark matter and dark energy—but rather the configuration in which we find our universe today. Remember that this already worked once, this kind of reasoning where we said, why is the universe in this kind of situation in which we find it? It worked with inflation. Inflation was able to successfully explain why the universe is homogeneous, and isotropic, and why it is very close to spatially flat, by invoking its dynamical mechanism based in physics located in the early universe.

But, it turns out that there is a deeper question that inflation doesn't quite answer, and that is very closely related to the question of the gas in the box. That's the question of the *entropy* of the universe. Our current universe has what we might call a medium-sized entropy. Entropy is basically telling you how disordered things are. If you have gas in a box which is stuck in one corner, that's a very low entropy. That's very highly ordered. Someone cleaned it up and put it there. If the gas is spread out all over the place, that's high entropy, disordered. We understand on the basis of physics that stuff in the universe likes to go from being low entropy to being high entropy. Disorder tends to increase just because there are many more ways of being disordered than being ordered. This is a deep feature of fundamental physics called the *second law of thermodynamics*. Things tend to become more and more disordered with time.

Our current universe is semi-ordered. In the far past, it was much more highly ordered and the entropy was much lower. In the far future, it will be much more disordered. The entropy will be much higher. The entropy is growing with time as the universe goes from the Big Bang to the cold heat death where everything is spread out and not radiating.

The question, then, is: Why? Why did the universe ever start out in some phase where the entropy was so low? You might, as many contemporary cosmologists do, think this has something to do with inflation, but in fact inflation is begging the question. Inflation imagines that a tiny patch of universe was dominated by dark energy in an even lower entropy state than the ordinary Big Bang. The truth is we don't yet know why the early universe had such a low entropy. It might, however, have something to do with dark energy.

Dark energy is telling us that, in the future of our universe, it will not be completely quiet. Dark energy means that even in empty space, there is some energy that keeps propelling the universe and keeps giving a little quantum jiggle to all the fields in the universe. If you wait long enough, much, much longer than the current age of the universe, it could be that those quantum jiggles come together in exactly the right configuration to make what is called a baby universe, to start with a little patch with incredibly high energy density ready to inflate and separate off from the universe that we know.

Furthermore, it could be that our universe is somebody else's baby, started off in some other universe that was cold and filled with nothing but dark energy. In other words, the dark energy that we've discovered might—we're just speculating here—play some role in helping us understand why the configuration of our current universe looks so unnatural, looks like it started in such a low entropy state. That's the kind of thing we're led to think about by thinking about dark matter and dark energy.

Finally, the cautionary tale is: How close are we to being done? In 1900, a lot of physicists thought that we were almost done with the laws of physics. We had a really good understanding on the basis of Newtonian mechanics of how matter and energy worked. There were just a few little things that were a tiny bit bothersome.

For example, if you calculated the lifetime of an atom, you found that it was an incredibly tiny fraction of a second. It should be the case that this electron that was zooming around the nucleus should collapse really, really quickly. Why was that number so much smaller than the observation said it was supposed to be? There was also an inconsistency in 1900 between successful theories of physics. Electromagnetism and Newtonian mechanics didn't quite have the same set of symmetries, and that was a puzzle. Finally, if you looked at the sky, you saw that some celestial bodies were not moving as they should. The orbit of Mercury was discrepant. It was moving in the wrong way, a different way than we predicted by Newtonian mechanics. As it turns out, all of these tiny discrepancies led to revolution. They led to quantum mechanics, to special relativity, and to general relativity. It could have been that you just cleaned up something here and there, or it could have been a completely different picture of the universe.

The same thing is true today. We have a number that is wrong. The number that is wrong is the vacuum energy. Our number is much smaller than it has any right to be. We observe a vacuum energy 120 orders of magnitude smaller than we predict. We have an inconsistency between successful theories. We have general relativity and quantum mechanics, both of which fit the data, but are mutually incompatible. Finally, we have celestial bodies moving in a way that doesn't make sense to us. We see things orbiting faster than they should. We attribute these to dark matter and dark energy.

It may be that we are almost done. It may be that a couple of things will fall into place and we will have a complete understanding of the fundamental laws of nature. Or, it could be that a revolution is in the offing similar to what we had 100 years ago. We don't know and that's the fun of science. The only guarantee is that the next 100 years of exploration will be equally interesting and surprising as the last 100 years have been.

Exponential Notation

Cosmology and particle physics involve some really big numbers, as well as some really small numbers. Scientists employ an elaborate system of notation and nomenclature to keep track of the quantities that define our universe.

The most important part of this system is **exponential notation**. It's awkward to write out numbers like a trillion, which is 1,000,000,000,000. So instead we just write 10^{12} (read as "ten to the twelfth power," or simply "ten to the twelve"), which represents a one followed by twelve zeroes. A thousand is 10^3, a million is 10^6, a billion is 10^9, and so on. The superscript that keeps track of the number of zeroes is known as the **exponent**, and it can equivalently be thought of as the number of 10s we multiply together to get the quantity under consideration: $10^4 = 10 \times 10 \times 10 \times 10$.

Since 10^0 means that we don't multiply any 10s, it should come as no surprise that $10^0 = 1$. Decreasing the exponent by one is equivalent to dividing the number by ten. Thus, a number like 10^{-1} stands for one-tenth (one divided by 10), or $10^{-1} = 0.1$. Similarly, $10^{-3} = 0.001$ is one-thousandth, and so forth.

We can express any number at all in exponential notation, just by multiplying an appropriate power of ten by some other number. So a number like 3,500 can equivalently be written as 35×10^2, or (more commonly) as 3.5×10^3. This kind of notation does more than just save ink; it quickly separates out the details about the precise number (the "3.5" part) from the rough idea of the size of the number (the "10^3" part). In cosmology, where numbers are very big and often hard to measure, the exponent is frequently the important part of the number.

Despite the elegance of exponential notation, physicists are people too, and they like to use words. So they've developed a shorthand vocabulary for exponents of various sizes (some of which are familiar from the metric system). "Milli-", for example, means "thousandth," or "times 10^{-3}." A millimeter is 10^{-3} meters, a milligram is 10^{-3} grams, and so on. Prefixes you might encounter in particle physics and cosmology include:

peta	tera	giga	mega	kilo	milli	micro	nano	pico	femto
10^{15}	10^{12}	10^9	10^6	10^3	10^{-3}	10^{-6}	10^{-9}	10^{-12}	10^{-15}

peta= 1,000,000,000,000,000   femto=.000000000000001

In abbreviations, positive powers of 10 are often written with capital letters, and negative powers with lower-case letters. A mega-electron volt, for example, is an MeV, while a milli-electron volt is an meV. The exception is "kilo-", which is abbreviated with a small "k," as in "km" for "kilometer."

Measuring the Universe

When we measure a physical quantity, we often do it by comparing it to something else. For example, we might observe that a football field is one hundred times longer than a yardstick. Many quantities can be mutually compared; for example, the height of a certain person might also be expressed as a multiple of the length of that yardstick. Other quantities cannot be compared; there's no way of relating the *weight* of that person to the *length* of the yardstick. Quantities that share a common standard of measure are said to have the same units. Extent in space, for example, is expressed in units of length.

The good news is that essentially every quantity in physics can be expressed using some combination of three basic units: **length, time**, and **mass**. The understanding of units, sometimes called "dimensional analysis," is actually very useful in physics. However, it is often the first thing that students are exposed to in high-school or college physics courses, which destroys their potential interest in the subject for the rest of their lives.

To make things just a bit more confusing, scientists tend to use different fundamental quantities to express the same kind of units, depending on the situation. We're used to that from our everyday lives; we regularly switch between seconds and minutes and hours and days and weeks (and so forth) to express different lengths of time. In fact, physicists are a bit more parsimonious than that; they tend to measure time using either seconds or years. One year is about 3.16×10^7 seconds (which many physicists remember as "approximately π times 10^7 seconds").

Length, of course, is measured in meters, although for mysterious reasons physicists often prefer to use centimeters (10^{-2} meters). In astrophysics, where lengths become quite large, we sometimes use kilometers, but very often revert to **light-years**: a light-year is the distance that light travels in one year, or 9.5×10^{12} kilometers (which is close enough to 10^{13} kilometers for most purposes). You will also frequently find **parsecs**; a parsec is about 3 light-years, or 3.1×10^{13} kilometers, to be precise. (It makes no sense to use parsecs when light-years are a perfectly good unit, but I don't make the rules.)

Mass is measured in grams, or sometimes kilograms. You might think that the non-metric unit for mass is the pound, but that's not quite true. Pounds are units of weight, which is not the same as mass—weight is actually a way of expressing the *force due to gravity* on a certain amount of mass. Of course, as long as we're comparing people and objects that are subject to the same gravitational force, as we usually are here on the surface of the Earth, weight is directly proportional to mass, which is why the concepts get confused. To make a long story short, an object that weighs one pound on the surface of the Earth will have a mass of 454 grams—a bit less than half a kilogram. (In orbit around the earth, the same object would be weightless, but its mass would still be 454 grams.)

Energy has units of velocity squared times mass, where velocity is length divided by time. For sake of convenience, we typically invent a unit to express energy directly, even though in principle we could stick to combinations of the existing units. One **erg** is defined as one gram times (one centimeter squared) divided by (one second squared). You will also come across **joules**; one joule is 10^7 ergs. In particle physics, the most common unit is the **electron volt** (eV), which is 1.6×10^{-12} ergs. The name comes from the amount of energy that an electron gains by being pushed through one volt of electric potential, but you don't have to remember that—it's just a convenient unit in which to measure energies. For example, the energy that binds an electron to a proton to make a single Hydrogen atom is 13.6 eV.

We don't need a separate set of units to measure temperature, because temperature is simply a way of expressing the average kinetic energy of particles in a substance. We nevertheless have several such units, because it's often convenient. Physicists usually use **Kelvins**, which are precisely proportional to energy—one K corresponds to 1.2×10^4 eV. If the particles are perfectly stationary, they have no kinetic energy, and the substance is said to be at **absolute zero**, the lowest possible point of the temperature scale. One Kelvin is exactly equal to one degree Celsius, although the two scales are related by an offset: zero Kelvin corresponds to –273 degrees Celsius. There are 1.8 familiar Fahrenheit degrees per Kelvin, and absolute zero is –460 degrees Fahrenheit.

There is one twist that seems subtle, but actually makes things much simpler once you get used to it. It makes perfect sense for us to use different units to measure distance and time—you wouldn't think of measuring someone's height using a stopwatch. But relativity teaches us that space and time are really two aspects of a single unified spacetime, so perhaps it's possible to use the same units for both quantities. Indeed it is, because there is one velocity (length divided by time) that has a special status in relativity—the **speed of light**, $c = 3.00 \times 10^{10}$ centimeters per second. Physicists therefore often find it convenient to measure distance and time in units where the speed of light is set equal to one. For example, if we measure distance in light-years and time in years, we have $c = 1.000$ light-years per year.

The real reason why it's incredibly convenient to use units in which $c = 1$ is because then the units of mass and energy are the same. Remember Einstein's famous formula for the rest energy of an object: $E = mc^2$. When $c = 1$, this is simply $E = m$! So you will very often hear physicists measure mass in what sounds like units of energy. One gram is equivalent to 5.6×10^{32} eV; the mass of the electron is 5.1×10^5 eV, and the mass of the proton is 9.4×10^8 eV (about 2,000 times the mass of the electron). This system is so convenient that it becomes second nature to particle physicists.

Nevertheless, we slowly-moving human beings tend to think as time and space as different things. For reference, here are some useful scales of time, length, and mass/energy. In some cases we have given very approximate values, either because the numbers are unknown (such as the mass of the hypothetical WIMP) or because there are a range of values (such as the mass of a typical galaxy). These are denoted by tildes ("~").

Time Scales

Planck time	5.4×10^{-44} sec
Inflationary Era	$\sim 10^{-35}$ sec
Electroweak Phase Transition	$\sim 10^{-12}$ sec
Big Bang Nucleosynthesis	~ 10 sec
Neutron lifetime (outside a nucleus)	880 sec
1 day	8.6×10^4 sec
1 year	3.2×10^7 sec
Human lifetime	$\sim 2 \times 10^9$ sec
Recombination	1.2×10^{13} sec $= 3.7 \times 10^5$ y
Age of the universe	4.4×10^{17} sec $= 1.4 \times 10^{10}$ y

Length Scales

Planck length	1.6×10^{-33} cm	
Proton size	$\sim 10^{-13}$ cm	
Atomic size	$\sim 10^{-8}$ cm	
Human being	~ 200 cm	
Earth radius	6.4×10^8 cm	
Sun radius	7.0×10^{10} cm	
Earth-Sun distance	1.5×10^{13} cm	$= 4.9 \times 10^{-6}$ pc
Distance to nearest star	4.1×10^{18} cm	$= 1.3$ pc
Size of typical galaxy	$\sim 10^{23}$ cm	$\sim 10^5$ pc
Hubble radius	1.3×10^{28} cm	$= 4.2 \times 10^9$ pc

Energy/Mass Scales

Axion mass (hypothesized)	$\sim 10^{-4}$ eV	
Neutrino masses	$\sim 10^{-3}$ eV	
Atomic binding energies	~ 1 eV	$\sim 2 \times 10^{-33}$ g
Electron mass	0.51 MeV	
Nuclear binding energies	~ 10 MeV	
Proton mass	938 MeV	
Neutron mass	940 MeV	
WIMP mass (hypothesized)	~ 100 GeV	
Planck mass	$\sim 10^{19}$ GeV	$\sim 10^{-5}$ g
Human being	$\sim 4 \times 10^{28}$ GeV	$\sim 7 \times 10^4$ g
Earth mass	3.4×10^{51} GeV	$= 6.0 \times 10^{27}$ g
Sun mass	1.1×10^{57} GeV	$= 2.0 \times 10^{33}$ g
Typical galaxy mass	$\sim 10^{69}$ GeV	$\sim 10^{45}$ g
Energy within observable universe	$\sim 10^{80}$ GeV	$\sim 10^{56}$ g

Timeline

The following is a list of important events in the evolution of the universe from the earliest times to the present. We have listed the *time* after the Big Bang, even though we don't know what happens at that moment itself. The *size* of the universe is a relative measure, given (approximately) in terms of the size today. The *temperature* refers to the temperature of the photon background, as well as any kind of particles that are in equilibrium with the photons.

The Planck era	
time = 10^{-43} sec size = 10^{-30} temperature = 10^{32} Kelvin	This is the era during which quantum gravity is important, and we don't really know what happens. We can't get any closer to the Big Bang at $t = 0$ and say anything with confidence.
Inflation?	
time = 10^{-35} sec size = 10^{-26} temperature = 10^{28} Kelvin	Theorists have speculated that the very early universe underwent a period of accelerated expansion, known as *inflation*, during which the energy density was temporarily dominated by a form of dark energy at an ultra-high-energy scale. This is a speculative theory, but one that has so far been consistent with observations.
Electroweak phase transition	
time = 10^{-12} sec size = 10^{-15} temperature = 10^{15} Kelvin	At high temperatures, electromagnetism is unified with the weak interactions. This is the temperature below which they become distinct.

Quark-gluon phase transition	
time = 10^{-6} sec size = 10^{-12} temperature = 10^{12} Kelvin	Above this temperature, quarks and gluons could exist as independent particles; afterwards, they become bound into the protons and neutrons we see today.
Primordial nucleosynthesis	
time = 10 sec size = 10^{-9} temperature = 10^9 Kelvin	The universe cools to a point at which protons and neutrons can combine to form light atomic nuclei, primarily helium, deuterium, and lithium. This is the earliest period about which we can test our understanding of cosmology against observational data.
Recombination	
time = 3.7×10^5 years size = 10^{-3} temperature = 3×10^3 Kelvin	The universe cools to a point at which electrons can combine with nuclei to form atoms, and the universe becomes transparent. Radiation in the cosmic microwave background is a snapshot of this era; tiny fluctuations in temperature, known as *anisotropies*, reveal a great deal about the early universe and its subsequent evolution.
The dark ages	
time = 10^8 years size = 10^{-1} temperature = 30 Kelvin	Small ripples in the density of matter gradually assemble into stars and galaxies.
Sun and Earth form	
time = 9×10^9 years size = 0.5 temperature = 6 Kelvin	From the existence of heavy elements in the Solar System, we know that the Sun is a second-generation star, formed about 5 billion years ago.

Today	
time = 13.7×10^9 years size = 1 temperature = 2.74 Kelvin	At least 100 billion galaxies shine within the observable universe. Dark energy is beginning to dominate over matter.

The Standard Model of Particle Physics

This is a list of the elementary particles in the Standard Model, including the one that has not yet been experimentally discovered (the Higgs boson).

First we group the elementary fermions—the quarks and leptons—into the three "families" or "generations" that particle physicists speak of based on charge and mass.

Families/Generations of Elementary Fermions (Quarks and Leptons)

	Family/Generation			
	1st	2nd	3rd	charge
Quarks	up	charm	top	+2/3
	down	strange	bottom	−1/3
Leptons	electron	muon	tau	−1
	electron neutrino	muon neutrino	tau neutrino	0

The next two tables give more detail about properties of elementary fermions, followed by similar information about elementary bosons. We list the mass of each particle, its spin, electrical charge, quark number (Q) and lepton number (L). The masses of the neutrinos are simply listed as "small," because they have not been accurately measured. Quark masses are also highly uncertain, and the values here are approximate.

Elementary Fermions (Quarks and Leptons)

Quark	Mass	Spin	Charge	Q, L
up (e)	3 MeV	1/2	+2/3	1, 0
down (d)	6 MeV	1/2	−1/3	1, 0
charm (c)	1.2 GeV	1/2	+2/3	1, 0
strange (s)	100 MeV	1/2	−1/3	1, 0
top (t)	171 GeV	1/2	+2/3	1, 0
bottom (b)	4.2 GeV	1/2	−1/3	1, 0

Lepton	Mass	Spin	Charge	Q, L
electron (e^-)	0.511 MeV	1/2	-1	0, 1
electron neutrino (ν_e)	small	1/2	0	0, 1
muon (μ)	106 MeV	1/2	-1	0, 1
muon neutrino (ν_μ)	small	1/2	0	0, 1
tau (τ)	1.8 GeV	1/2	-1	0, 1
tau neutrino (ν_τ)	small	1/2	0	0, 1

Next we list the five types of bosons in the Standard Model. Four of these correspond to the four fundamental forces of nature: the electromagnetic force (the photon); the strong force (eight types of gluons); the weak nuclear force (three bosons: the W^+, the W^-, and the Z bosons); and gravity (the hypothetical graviton). A fifth type of boson (the Higgs boson) is hypothesized to exist in order to explain why fundamental particles can have mass.

Elementary Bosons (force-carrying particles)

Boson	Mass	Spin	Charge	Q, L
photon (γ)	0	1	0	0, 0
gluons (g)	0	1	0	0, 0
W^+	80 GeV	1	$+1$	0, 0
W^-	80 GeV	1	-1	0, 0
Z^0	91 GeV	1	0	0, 0
graviton (G)	0*	2	0	0, 0
Higgs (h)	100–200 GeV*	0	0	0, 0

*hypothesized values only for the mass of the graviton and of the Higgs boson

The final table lists some of the hadrons, which are composite particles. Hadrons consisting of three quarks are known as baryons: for example, the proton consists of two up quarks and one down, while the neutron consists of two down quarks and one up. Hadrons consisting of a quark-antiquark pair are known as mesons; for example, pions are quark-antiquark pairs.

Selected Hadrons (Baryons and Mesons)

Baryon	Mass	Spin	Charge	Q, L
proton (p)	938.3 MeV	1/2	+1	3, 0
neutron (n)	939.6 MeV	1/2	0	3, 0

Meson	Mass	Spin	Charge	Q, L
positive pion (π^+)	140 MeV	0	+1	0, 0
negative pion (π^-)	140 MeV	0	−1	0, 0
neutral pion (π^0)	135 MeV	0	0	0, 0

The above tables do not include antiparticles, except for the W^+/ W^- and π^+/π^- (which are each other's antiparticles) and those particles that are their own antiparticles (γ, h, Z^0, G, π^0).

Note: as discussed in the "Measuring the Universe" appendix, particle masses are traditionally given in terms of electron volts (eV), a unit of energy. That's because physicists implicitly multiply by the speed of light squared, using $E = mc^2$, to convert mass into energy. A *mass* of 1 electron volt is equivalent to about 2×10^{-33} grams. Common abbreviations include keV (kilo-eV, or 1,000 eV), MeV (mega-eV, or 1 million eV), and GeV (giga-eV, or 1 billion eV).

Glossary

acceleration of the universe: Distant galaxies have an apparent recession velocity given by Hubble's law. If the velocity is increasing with time, we say that the universe is accelerating. In general relativity, the universe can accelerate only if it is dominated by a persistent negative-pressure component, known as *dark energy*.

anisotropy: *Isotropy* means that things are exactly the same in every direction, and *anisotropy* means that they are slightly different. Our universe is approximately isotropic on large scales but not exactly so; in particular, the tiny anisotropies in the temperature of the cosmic microwave background arise from density perturbations that later grow into galaxies and large-scale structure.

annihilation: What is said to happen when a particle and its antiparticle collide and convert into some other kind of particles.

anthropic principle: The idea that certain features of the universe can be explained by the simple observations that if they were not that way, life could not exist, and we wouldn't be around to measure them.

antimatter: Every particle of matter also has an associated particle of antimatter. The conserved quantities associated with antiparticles (electric charge, quark number, baryon number) are opposite to those of the original particles, but the mass is exactly equal. When a particle has no conserved quantities, such as a photon or Z boson, it is its own antiparticle. When particles and antiparticles collide, they typically annihilate into radiation; for this reason, antimatter is not a good candidate for dark matter. The reason why there is much more matter than antimatter in the universe is an unsolved puzzle; see **baryogenesis**.

atom: The combination of a nucleus and its surrounding electrons. The original use of *atom* by the ancient Greeks referred to the smallest indivisible unit of matter, which is obviously inappropriate for our modern-day atoms. That's simply because physicists who discovered modern atoms were too hasty in attributing to them the property of being indivisible. It remains true, however, that atoms are the smallest units of individual chemical elements.

axion: A hypothetical boson that is a promising dark matter candidate. Even though axions have very low masses, they are produced at a very low temperature in the early universe, so they would count as *cold* dark matter.

baryogenesis: The hypothetical process in the early universe in which baryons were created preferentially over antibaryons, leading to the preponderance of matter over antimatter in our current universe.

baryon: Any kind of particle with a positive (nonzero) quark number. Protons and neutrons, with three quarks each, are baryons. Antiprotons and antineutrons are antibaryons, although sometimes both types of particles are lumped together as *baryons*. Pions and other mesons, consisting of one quark and one antiquark, have quark number 0; they are hadrons, but not baryons.

baryon number: Quark number divided by 3. Because quarks are confined into colorless combinations, all physical particles have quark numbers that are divisible by 3, which is why baryon number is convenient. Protons and neutrons each have baryon number 1.

Big Bang: The term *Big Bang* has different meanings. It can refer to the hypothetical moment of infinite temperature and density in the early universe, although it is conventionally believed that such a singularity is an artifact of our incomplete understanding of physics in those conditions. Therefore, *Big Bang* is sometimes used to refer to the very earliest moments in the history of the universe, whatever they may have been. Also, the phrase *Big Bang model* is often used to refer to the entire picture of a nearly uniform universe expanding from an initial hot, dense state to its current configuration billions of years later.

Big Crunch: A future singularity of a universe that recollapses. Our universe is accelerating, and there is no evidence that it ever will collapse; however, without a complete understanding of the dark energy, we can't be sure.

Big Rip: A future singularity of a universe that accelerates (not just expands) at an ever-increasing rate. At the Big Rip, the expansion of space tears apart objects, all the way down to individual atoms. We have no reason to believe that our universe will experience a Big Rip, but it is a theoretical possibility.

black hole: A region of spacetime in which the force of gravity is so strong that light itself cannot escape. Particles inside a black hole move inevitably toward the singularity. Black holes can be formed from the collapse of individual stars or, in the form of *supermassive* black holes (greater than a million solar masses), at the centers of galaxies. Such black holes came from ordinary matter so they don't count as dark matter. See also **massive compact halo object**.

blackbody radiation: Any object at a uniform temperature emits photons with a very specific distribution of energies. This is blackbody radiation; in cosmology, the most famous example is the cosmic microwave background, left over from the early universe.

boson: A force-carrying particle, as opposed to a matter particle (fermion). Bosons can be piled on top of each other without limit. Examples include photons, gluons, gravitons, weak bosons, and the Higgs boson. The spin of a boson is always an integer, such as 0, 1, 2, and so on.

bottom quark: An elementary particle, one of the six types of quarks. It is the second-heaviest quark after the top.

brane: Any type of fundamental object that is extended in one or more dimensions. Branes are like particles, except that they can be one-dimensional or two-dimensional (or even more if extra dimensions exist), rather than simply zero-dimensional. A one-dimensional brane is a string, and a two-dimensional brane is a wall (or membrane, from which the term is derived).

brown dwarf star: See **massive compact halo object**.

causal contact: Two particles are said to be in *causal contact* if signals from a common point in the past can reach both of them while traveling at or below the speed of light. Because the universe has a finite age, events can exist that are outside of causal contact. A period of inflation in the early universe can explain how widely separated points can be in causal contact.

Cepheid variable: A type of pulsating star. The period of pulsation is directly related to the brightness of the star, allowing Cepheids to be used as standard candles.

Chandrasekhar limit: The maximum mass of a white dwarf star, about 1.4 times the mass of the Sun. Above this mass, the gravitational pull becomes too great, and the star must collapse to a neutron star or black hole.

charm quark: An elementary particle; one of the six types of quarks. It is midway in mass between the strange quark and the bottom quark.

classical mechanics: The framework for physics developed by Isaac Newton. In classical mechanics, systems evolve deterministically and can be observed to arbitrarily high precision. Classical mechanics was superseded by quantum mechanics.

cluster: A collection of galaxies orbiting each other under the influence of their mutual gravitational fields.

coincidence scandal: The fact that the energy density in matter and dark energy are roughly comparable in the current universe when we are around to observe them. Dark energy has an approximately constant density, while matter dilutes rapidly as the universe expands; therefore, these two quantities change dramatically with respect to each other. In the early universe, the density of dark energy was much less than that of matter, while in the far future, the situation will be reversed. The fact that we currently live at the point of approximate balance seems to be a coincidence, although it's possible that we will eventually find a deep explanation.

cold dark matter: A kind of dark matter in which the individual particles are moving slowly compared to the speed of light (and always have been). *Hot dark matter*, in contrast, would have rapidly moving particles. Cold dark matter seems to be a much better fit to the data, because hot particles would not be able to collect into galaxies and clusters.

color: Quarks come in three colors, conventionally thought of as red, green, and blue. Antiquarks have anti-colors (anti-red and so on). The color of a quark is the strong-interaction equivalent of the electric charge. A remarkable property of the strong interaction is the confinement of quarks into colorless combinations: either three quarks, one each of red, green, and blue, or one quark and one antiquark.

concordance cosmology: The basic modern cosmological picture, in which the energy density of the universe is approximately 5 percent ordinary matter, 25 percent dark matter, and 70 percent dark energy.

confinement: The strong interaction confines individual quarks (and gluons) into colorless combinations, such as baryons and mesons. The amount of energy that would be required to separate an individual quark is greater than the amount needed to create a new quark/antiquark pair. Therefore, it is impossible to observe quarks all by themselves. At very high energies or temperatures, however, confinement is violated; the early universe was filled with free quarks and gluons.

conserved quantity: A property of individual particles or collections of particles that remains unchanged during any interaction, even as particles change their identities. Electric charge is a classic example. In the Standard Model, other examples include quark number and lepton number; it is possible, however, that unknown physics will eventually be found that violates these quantities.

cosmic microwave background (CMB): Leftover radiation from the Big Bang. The early universe was hot and dense, filled with blackbody radiation at a high temperature. It was also opaque, as photons frequently collided with free electrons. At the moment of recombination, when electrons joined up with protons to make hydrogen atoms, the universe became transparent, and the photons streamed freely through space. They were subsequently redshifted to the microwave portion of the electromagnetic spectrum at an effective temperature of about 2.7 Kelvin. We observe that radiation today as the cosmic microwave background. Tiny deviations in the CMB temperature from place to place provide a snapshot of the universe when it was about 380,000 years old. Those anisotropies arise from fluctuations in the matter density, which have since grown into galaxies and large-scale structure.

cosmological constant: Another name for vacuum energy. Originally introduced by Einstein to help produce static solutions to his equation for the size of the universe but later repudiated by him when Hubble discovered that the universe is expanding. The cosmological constant is now the leading candidate for dark energy.

cosmological constant problem: The puzzle associated with the magnitude of the vacuum energy. Simple arguments lead us to expect a value that is 10^{120} times bigger than what is actually observed. The explanation for this discrepancy is still unknown.

cosmological principle: The idea that the universe is pretty much the same everywhere. In more technical terms, on sufficiently large scales, the universe is homogeneous and isotropic.

critical density: The density that would be required to make the universe spatially flat, as required by the Friedmann equation.

curvature of space: In general relativity, *spacetime* is a four-dimensional manifold, and the curvature of that manifold is what we interpret as gravity. In a homogeneous and isotropic universe, the curvature of *space* alone (which is a subset of the curvature of spacetime) is the same everywhere. It can be zero, in which case, space is flat, and traditional Euclidean geometry is applicable. It can be positive, in which case, space is a three-dimensional version of a sphere. Characteristics of positively curved spaces are that the interior angles of a triangle will always add up to greater than 180 degrees and initially parallel lines will eventually intersect. It can also be negative, analogous to the surface of a saddle. In a negatively curved space, interior angles of a triangle add up to less than 180 degrees and initially parallel lines eventually diverge.

dark energy: A smooth, persistent component of invisible energy, thought to make up about 70 percent of the current energy density of the universe. Dark energy is known to be smooth because it doesn't accumulate preferentially in galaxies and clusters. Its persistence—the feature that the density of dark energy remains approximately constant as the universe expands—imparts an accelerated expansion to the universe. Evidence for dark energy comes from observations of supernovae that indicate the universe is accelerating and from observations of the cosmic microwave background that measure the total amount of energy in the universe, which turns out to be greater than the energy density in matter (dark and ordinary). The precise nature of the dark energy is not currently understood; it could be a strictly constant vacuum energy or a gradually evolving quintessence field. It is also conceivable that the evidence for dark energy is misleading, and we are really seeing the effects of a departure from Einstein's theory of gravity on very large scales.

dark matter: An invisible, essentially collisionless component of matter that makes up about 25 percent of the energy density of the universe. Ordinary matter made of atoms is about 5 percent of the total energy density, so the dark matter is not simply ordinary matter that is dark; it's a different kind of particle. (Some references use the term *dark matter* to refer to both ordinary matter that is invisible and to this new particle, but that usage has become less frequent.) All candidates in the Standard Model of particle physics have been excluded, so the dark matter particle is something not yet observed in the laboratory.

decay: The spontaneous transformation of a single particle or nucleus into lighter particles. The neutron decays into a proton, an electron, and an electron anti-neutrino, with a lifetime of about 10 minutes. Particles that do not decay are called stable.

density fluctuations: On very large scales, the density of the universe is quite uniform but not perfectly so. From very early times, there were small differences in the amount of matter from place to place, known as *density fluctuations*. Under the influence of gravity, those tiny early fluctuations increased in magnitude as time passed, leading to the formation of galaxies and large-scale structure. Anisotropies in the temperature of the cosmic microwave background provide a precise probe of the properties of density fluctuations at early times.

deuterium: An isotope of hydrogen with 1 proton and 1 neutron in the nucleus.

Doppler effect: The perception on the part of an observer who is moving with respect to a source of light (or sound) that the wavelength is shifted away from that created at the source. If the observer is moving toward the source, the wavelength is shorter (a *blueshift*), while an observer moving away perceives a longer wavelength (a *redshift*). The cosmological redshift is related to the Doppler shift but is not quite the same thing; strictly speaking, it is not due to motions of the galaxies but to the expansion of space in between them.

down quark: An elementary particle; one of the six types of quarks. It is the second lightest quark after the up and, as a constituent of both protons and neutrons, plays an important role in ordinary matter.

Einstein's equation: The fundamental equation (first derived by Einstein) of general relativity. It describes how matter and energy give rise to the curvature of spacetime. In mathematical notation, it reads $R_{\mu\nu} - (1/2)Rg_{\mu\nu} = (8\pi G)T_{\mu\nu}$. On the left-hand side, $R_{\mu\nu} - (1/2)Rg_{\mu\nu}$ is the *Einstein tensor*, a four-by-four matrix characterizing the curvature of spacetime. On the right-hand side, $T_{\mu\nu}$ is the *energy-momentum tensor*, a four-by-four matrix characterizing the matter and energy sources. G is Newton's constant of gravitation.

electric charge: The conserved quantity that acts as a source for the electric field. In natural units, the charge of the electron is –1, the charge of the proton is +1, and the charge of the neutron is 0.

electromagnetism: A triumph of 19th-century physics was the unification of the theories of electricity and magnetism into a unified picture, known as *electromagnetism*. Modern physics views electric fields and magnetic fields as manifestations of a single underlying phenomenon.

electron: An elementary particle; the lightest charged lepton. Electrons are perfectly stable because they are the lightest charged particle of any sort and, therefore, have nothing to decay into. Atoms consist of electrons surrounding nuclei made of protons and neutrons. Anti-electrons are usually referred to as *positrons*.

electron neutrino: An elementary particle; the neutrino that interacts directly with the electron.

electron volt (eV): A unit of energy. Using $E = mc^2$, 1 electron volt is equivalent to about 2×10^{-33} grams of mass.

electroweak theory: At low energies, the electromagnetic force and the weak nuclear force are very different; most obviously, electromagnetism operates over very large ranges, while the weak force is very short-ranged. At sufficiently high energies, however, these two forces are unified into a single theory.

element: The kind of ordinary matter consisting of atoms with a specific number of protons in the nucleus. Hydrogen is the lightest element, with 1 proton in the nucleus; helium is the next lightest, with 2; and so on. Nuclei with equal numbers of protons but different numbers of neutrons are referred to as different *isotopes* of the same element.

energy density: The amount of energy per volume of space, measured (for example) in units of ergs per cubic centimeter.

equation-of-state parameter: The ratio of the pressure of a fluid to its energy density, often denoted w. The equation-of-state parameter is important in cosmology because it governs the rate at which the energy density evolves as the universe expands. For matter, $w = 0$; for radiation, $w = 1/3$; and for vacuum energy, $w = -1$. The value -1 corresponds to an energy density that doesn't change as the universe expands; if the equation-of-state parameter is slightly higher (such as -0.9), the density slowly decreases, while if it is slightly lower (such as -1.1), the density slowly increases.

equivalence principle: An idea expressing the equivalence of gravity and uniform acceleration. Unlike other forces, gravitation acts on all particles in the same way—any two objects, dropped in a gravitational field, will fall along the same trajectory. Therefore, by doing experiments inside a sealed room, it is impossible to determine whether the room is sitting in a gravitational field or accelerating through empty space. Einstein was inspired by the equivalence principle to suggest that gravity could be thought of as a feature of the curvature of spacetime, rather than as a separate force propagating through spacetime.

erg: A unit of energy equal to about 600 GeV. Using $E = mc^2$, 1 erg is equivalent to a mass of about 2×10^{-21} grams.

extra dimension: An additional, hidden dimension of space beyond the three that we readily experience. The idea of extra dimensions goes back to Kaluza and Klein, not long after Einstein proposed general relativity. Obviously, such dimensions must be hidden somehow from our view; the conventional method is to imagine that they are compact, with sizes far below what we can observe (perhaps at the Planck scale). Modern ideas use branes to suggest ways that extra dimensions could be substantially larger than that: If all the particles of the Standard Model were confined to a brane, the extra dimensions would be invisible simply because we couldn't get there.

family: See **generation**.

fermion: A matter particle, as opposed to a force particle (boson). Fermions take up space and can't be piled on top of each other. Examples include all varieties of quarks and leptons. The spin of a fermion is always a half-integer (1/2, 3/2, 5/2, and so on).

Feynman diagram: A graphical representation of an interaction between elementary particles. Physicists use Feynman diagrams both as a way to keep track of what processes can occur and as a calculational tool for determining the frequency of such processes.

fifth force: We usually think of four forces in nature: gravity, electromagnetism, the strong nuclear force, and the weak nuclear force. (There is also a force from the Higgs boson, but it has not yet been detected.) Any additional "fifth" force, especially one that operates over long distances, would be evidence for a new light boson, which might be related to the hypothetical quintessence field. One way to discover such a force is to find that objects of different compositions fall at slightly different rates.

flatness problem: Observations indicate that the spatial curvature of our universe is very close to zero. But in a universe with only matter and radiation (no dark energy), spatial curvature increases with time. The puzzle about why it started so small is the *flatness problem*. **Inflation** is a proposed solution.

flavor: The six different types of quarks are sometimes referred to as *flavors*.

field: A quantity that consists of a number (or set of numbers) at every point in spacetime. The electric field, for example, consists of a vector at every point in spacetime. Modern physics describes all physical particles as vibrations of different kinds of fields—there is an electron field, a gluon field, and so forth. The fact that we observe localized particles rather than smooth fields is a consequence of quantum mechanics.

Friedmann equation: The fundamental equation governing the expansion of the universe in general relativity. The Friedmann equation is a special case of Einstein's equation, applied to a homogeneous and isotropic universe. It relates the Hubble parameter, H, to the energy density, ρ (rho), and the spatial curvature, K. In mathematical notation, it takes the form $(8\pi G/3)\rho = H^2 + K$, where G is Newton's constant of gravitation.

galaxy: A collection of stars, gas, and dark matter, orbiting under their mutual gravitational pull. The galaxy we live in is the Milky Way.

gamma rays: Ultra-high-energy electromagnetic radiation, with wavelengths between 10^{-14} and 10^{-11} meters.

gauge boson: A force-carrying particle of the Standard Model. Examples include photons, gluons, and the W and Z bosons.

general relativity (GR): Einstein's theory of gravitation. The basis of GR is the idea that gravity is the curvature of spacetime. Sources of gravity—energy, mass, and momentum—are related to spacetime curvature by Einstein's equation. In this curved spacetime background, particles do their best to travel on straight lines, but there are no straight lines because of the curvature. We interpret the resulting motions as being due to the force of gravity.

generation: Both the quarks and the leptons of the Standard Model come in three generations. The lepton generations consist of one charged lepton and its associated neutrino. The lightest generation consists of the electron and the electron neutrino; the next heaviest is the muon and the muon neutrino; and the heaviest is the tau and the tau neutrino. The lightest generation of quarks consists of the up and down quarks; the next generation is the charm and strange quarks; and the heaviest generation is the top and bottom quarks. Generations are sometimes known as *families*. The reason why the same pattern of particles is repeated three times is still not understood.

GeV: Giga-electron volt (1 billion eV). See "Measuring the Universe" appendix.

gluon: The boson that carries the strong nuclear force. There are actually eight different kinds of gluons, representing different combinations of colors and anti-colors. Like quarks, gluons are confined inside hadrons; thus, we don't see individual gluons.

gravitational lensing: The deflection of light by a gravitational field. By measuring the amount of deflection, astronomers can infer the amount of mass in a system.

gravitational wave (gravitational radiation): Oscillating gravitational fields propagating through space at the speed of light. Just as electromagnetic radiation can be caused by an oscillating electric charge, gravitational radiation can be caused by oscillating masses, such as orbiting black holes. Gravitational waves are generally very weak and have not yet been directly observed, although the search for them is actively going on.

gravitino: Hypothetical supersymmetric partner of the (hypothetical) graviton. Given that the Higgs is a boson, the gravitino is a fermion.

graviton: The force-carrying particle for gravity. Individual gravitons have never been observed because of the weakness of the gravitational force. Given that we do not have a consistent theory of quantum gravity, we cannot even be sure that there are such things as gravitons. However, the existence of gravitons seems to be a robust prediction of the most basic features of quantum mechanics and general relativity.

hadron: Any individual particle comprised of quarks and gluons. Hadrons with positive quark number are *baryons*, and those with negative quark number are *antibaryons*. Mesons, consisting of a quark and an antiquark, are hadrons even though they are not baryons.

Heisenberg uncertainty principle: In quantum mechanics, the uncertainty principle places a fundamental limit on the precision with which observers can simultaneously measure the position and velocity of a particle. If the position is measured very precisely, the velocity will be very poorly determined and vice versa. In quantum field theory, the uncertainty principle forbids us from precisely pinning down the state of a field; this uncertainty implies the existence of vacuum fluctuations, in which virtual particles appear and disappear in empty space.

helium: The element with 2 protons. The most common isotope is helium-4, with 2 neutrons.

Higgs boson: A hypothetical boson; the only particle in the Standard Model of Particle Physics that has not yet been detected. The existence of the Higgs boson is predicted by the Higgs mechanism, the process that breaks electroweak symmetry and gives masses to elementary fermions. The Higgs is unstable, with a very short lifetime, and is therefore not a candidate for dark matter.

Higgsino: Hypothetical supersymmetric partner of the (hypothetical) Higgs boson. Given that the Higgs is a boson, the Higgsino is a fermion.

©2007 The Teaching Company.

homogeneity: Having the same properties at every point. Homogeneity of the universe means that the density of matter and energy is (approximately) the same on very large scales.

horizon: Because of the finite speed of light and the finite time since the Big Bang, we have access only to a finite region of the universe; other points are out of causal contact with us. The horizon is the point past which no signal can reach us.

horizon problem: In conventional cosmology, if the early universe were dominated exclusively by matter and radiation, widely separated points on the cosmic microwave background were outside each other's horizons. Nevertheless, they appear to be very close in temperature. The horizon problem is simply the puzzle of how points that were never in causal contact could have the same temperature. The most promising solution to this problem is offered by inflation.

hot dark matter: Dark matter that is moving near the speed of light or, at least, was moving near the speed of light in the early days of structure formation. Such particles would not be able to settle into galaxies as efficiently as cold dark matter particles do; the hot dark matter model is, therefore, disfavored.

Hubble constant (Hubble parameter): The ratio of the apparent speed of galaxies to their distances, as encompassed by Hubble's law.

Hubble distance: A rough guide to the size of the observable universe, given by the speed of light divided by the Hubble constant.

Hubble's law: The proportionality between the distance and the apparent recession velocities of galaxies is known as *Hubble's law* ($v = Hd$): The further away a galaxy is, the faster it appears to be receding. (We say *appears* because it's really that space is expanding, not that galaxies are moving through space.) The ratio of the speed to the distance is *Hubble's constant*. Hubble's law doesn't apply exactly to nearby galaxies (where individual motions are important) or to galaxies that are very far away (where the change in the expansion rate over time becomes important).

hydrogen: The element with 1 proton. The most common isotope, it has no neutrons at all; the atomic form consists of just a single proton and a single electron. The isotope deuterium (sometimes called *heavy* *hydrogen*) contains a single neutron in the nucleus as well; the unstable isotope tritium contains 2 neutrons.

inflation: A hypothetical period of rapidly accelerated expansion in the very early universe. Inflationary expansion smoothes out space, potentially solving the horizon and flatness problems.

inverse-square law (Newton's law of gravitation): The rule according to which the strength of the gravitational force between two objects decreases as the square of the distance between them increases. Although Newton's understanding of gravitation has been superseded by general relativity, the inverse-square law still holds when gravity is not too strong.

isotope: A nucleus with a specific number of protons and neutrons. Deuterium (1 proton and 1 neutron) and tritium (1 proton and 2 neutrons) are both isotopes of hydrogen, whose most common isotope consists of just a single proton.

isotropy: Looking the same in every direction. The large-scale universe, especially the temperature of the cosmic microwave background, looks very isotropic, although not exactly so. Combining observed isotropy with the cosmological principle implies that the universe is homogeneous.

Kelvin: A unit of temperature; 1 Kelvin is the same as 1° Celsius and is equal to 1.8° Fahrenheit. Zero on the Kelvin scale corresponds to −273° Celsius or −460° Fahrenheit. It is the lowest possible physical temperature, known as *absolute zero*.

keV: Kilo-electron volt (1,000 eV).

kinetic energy: Energy of motion. Kinetic energy stands in contrast to rest energy, which objects have even when they are not moving.

landscape: The hypothetical set of possible phases in which spacetime can find itself, according to string theory. The landscape may have as many as 10^{500} different phases; the precise number is quite uncertain at this time. If inflation operates to make all these phases actually exist somewhere, we have a multiverse.

lepton: Standard Model fermion that does *not* feel the strong nuclear force. The charged fermions include the electron, the muon, and the tau; each has a corresponding neutrino, which is also a lepton.

lepton number: A conserved quantity given by the total number of leptons minus the total number of antileptons. Electrons, muons, taus, and neutrinos all have lepton number 1; positrons and other antileptons have lepton number −1.

lifetime: The average time an unstable particle will last until it decays.

light cone: The set of points in spacetime that can be reached from a given event by traveling at the speed of light. The *future light cone* is the set of points that can be reached by traveling from the fixed event, while the *past light cone* is the set of points from which the fixed event can be reached. Because nothing can travel faster than the speed of light, the only points that can ever be influenced by some event are those within that event's future light cone. Similarly, the only points that can ever influence that event are those that are within its past light cone.

light-year: Distance traveled by light in one year, about 9×10^{12} kilometers.

lightest supersymmetric particle (LSP): In supersymmetry, the hypothetical lightest superpartner of the known Standard Model particles. Because supersymmetric particles have not yet been experimentally discovered, we don't yet know which superparticle is the LSP. If it is neutral and stable, the LSP makes an ideal WIMP candidate for dark matter.

lithium: The element with 3 protons. The most common isotope is lithium-7, with 4 neutrons.

mass: In relativity, the *mass* of a particle is simply a measure of the energy it has when it is at rest. Using $E = mc^2$, physicists convert readily between energy and mass and will often make such statements as "The mass of the electron is 511,000 electron volts," even though the electron volt is a unit of energy. What they mean is, if you multiply the mass of the electron by the speed of light squared, you get an energy equal to that many electron volts.

massive compact halo object (MACHO): A dim star or planet-like object. MACHOs can range in size from brown dwarf stars to massive planets or could even be black holes. Although they are dark, they count as ordinary matter rather than as part of the dark matter.

matter: This term has different meanings in different contexts. In particle physics, we distinguish *matter particles* (fermions, which take up space) from *force particles* (bosons, which can pile on top of each other). In cosmology, *matter* refers to any collection of slowly moving particles (where *slow* is a comparison with the speed of light). We then have *ordinary matter* or *dark matter*. The energy of slowly moving particles is approximately equal to their rest energy, which is a constant. Therefore, the energy density in matter is proportional to the number density, which declines as the volume goes up.

meson: A composite bosonic particle consisting of one quark and one antiquark. Mesons are hadrons (given that they are made of quarks), but they are not hadrons (given that their total quark number is 0).

MeV: Mega-electron volt (1 million eV). See the "Measuring the Universe" appendix.

microlensing: Minor occurrences of gravitational lensing, when a dim star or planet passes in front of a brighter star. The gravitational field of the lensing object deflects away some of the brighter object's light, leading to a temporary dimming in brightness, even if no distortion in shape or position is observed.

Milky Way: The galaxy in which we live, consisting of about 100 billion stars.

modified Newtonian dynamics (MOND): A model of gravity in which the gravitational force falls off more slowly at large distances (small accelerations) than in conventional Newtonian gravity. Originally inspired by the dynamics of spiral galaxies, MOND does a pretty good job at fitting individual galaxies but fails when applied to clusters and has nothing to say about the expansion of the universe.

multiverse: The hypothetical concept of unobservable regions of space where conditions are potentially very different from those we see in our observable universe. The "other universes" may be truly separate from ours or simply too far away to be accessible. The multiverse, with its wide variety of conditions, plays an important role in the anthropic principle.

muon: An elementary particle; a lepton similar to the electron but heavier and unstable. The discovery of the muon in 1936 was the first evidence of the second generation of elementary particles.

muon neutrino: An elementary particle; the neutrino that interacts directly with the muon.

nebula (pl.: nebulae): Originally, a nebula was simply a celestial object that appeared fuzzy in telescopes, as opposed to pointlike stars. It was later realized that some nebulae are collections of gas and dust within our galaxy, while others are separate galaxies in their own right.

neutron: A composite fermionic particle consisting of two down quarks and one up quark. Neutrons are both hadrons and baryons. The neutron has electric charge 0, quark number +3, and lepton number 0. It decays into a proton, an electron, and an electron anti-neutrino, with a lifetime of about 10 minutes.

neutrino: An elementary fermionic particle. Neutrinos are uncharged leptons and come in three flavors: the electron neutrino, the muon neutrino, and the tau neutrino. Electron neutrinos interact directly with electrons but not with muons or taus, and so forth. Although they are much less massive than any of the quarks or charged leptons, it was recently discovered that neutrinos do have small nonzero masses.

neutron star: A type of compact star consisting of (almost) nothing but neutrons. Neutron stars are formed when the mass of a white dwarf is greater than the Chandrasekhar limit, and gravity forces the electrons and protons to combine into neutrons.

Newton's law of gravitation: See **inverse-square law**.

nucleon: Protons and neutrons are collectively known as *nucleons* because they make up the atomic nucleus.

nucleosynthesis: The process by which heavy elements are formed from lighter ones. Primordial nucleosynthesis occurred when the universe was a few minutes old, and protons and neutrons were converted into hydrogen, deuterium, helium, and lithium. Heavier elements are formed in stars and supernova explosions in the later universe.

nucleus (pl.: nuclei): A bound collection of protons and neutrons. Nuclei have a positive electric charge (from their protons) and generally capture electrons to form atoms.

ordinary matter: Matter consisting of particles in the Standard Model of Particle Physics—which includes every kind of matter ever directly observed (so far). In the current universe, about 5 percent of the total density takes the form of ordinary matter. It is sometimes referred to as *baryonic matter*, because the large majority of the mass density of ordinary matter comes in the form of baryons (protons and neutrons in atomic nuclei).

parallax: The change in apparent position of a distant (but not too distant) object when you look at it from two different places. The further away something is, the smaller its parallax; this effect, therefore, provides an excellent way to determine the distances to astronomical objects. Unfortunately, the parallax of objects outside our galaxy is too small to be observed, and we must rely on less direct distance indicators.

Pauli exclusion principle: The impossibility of putting two fermions into the same state. It is this property of fermions that makes them matter particles—they take up space. Bosons, in contrast, do not obey the exclusion principle and can be squeezed together without limit.

phantom energy: A hypothetical form of dark energy, in which the energy density is gradually increasing as the universe expands. If such an increase continues in perpetuity, the future of the universe can experience a Big Rip.

photino: Hypothetical supersymmetric partner of the photon. Given that the photon is a boson, the photino is a fermion.

photon: The bosonic particle that mediates the electromagnetic force. An electromagnetic wave or field consists of a condensate of a large number of photons. Photons interact directly with any kind of particle that carries electric charge.

pion: A composite meson consisting of one quark and one antiquark. There are three types of pions: positively charged, negatively charged, and neutral. Pions may be thought of as the particles carrying the force that keeps protons and neutrons together inside the nucleus.

Planck scale: The scale at which quantum gravity is presumed to become important. In different contexts, *Planck scale* can refer to the Planck energy (roughly 10^{28} eV), the Planck mass (roughly 10^{-5} g), the Planck length (roughly 10^{-33} cm), or the Planck time (roughly 10^{-43} s).

positron: The antiparticle of the electron.

pressure: If you have a gas in a box, the force per unit area exerted by the gas on the sides of the box is the pressure. In the middle of the box, the net force is 0, because it is pushing equally in all directions, but the pressure is still there. In cosmology, there is no box, but the gravitational effect of the pressure of different cosmological fluids is very important. Matter (slowly moving particles) has approximately 0 pressure; radiation (particles moving near or at the speed of light) has a pressure that is 1/3 of its energy density; and vacuum energy has a pressure that is exactly the opposite of its energy density ($p = -\rho$). Thus, if the vacuum energy is positive, its associated pressure is negative, sometimes known as *tension*.

proton: A composite fermionic particle consisting of two up quarks and one down quark. Protons are both hadrons and baryons. The proton has electric charge +1, quark number +3, and lepton number 0. As the lightest particle with a nonzero quark number, it appears to be stable, although it is conceivable that protons have a lifetime that is very large but not infinite (if quark number is not precisely conserved).

quantum chromodynamics (QCD): The theory that explains the strong nuclear force in terms of quarks and gluons. In QCD, the quantity that plays a role analogous to electric charge is color, which gives rise to the name *chromodynamics*. Just as photons couple to electric charge, gluons couple to color. According to QCD, color is confined inside hadrons, so that observable particles are always colorless combinations of quarks.

quantum electrodynamics (QED): The quantum theory of electricity and magnetism. QED describes electromagnetic forces as arising from the exchange of photons, which couple to electric charge.

quantum field theory: As you might guess, the quantum-mechanical theory of fields. In classical field theory, fields stretching through space obey deterministic equations of motion and can be precisely measured. In quantum field theory, the state of the field is contained in a wavefunction that describes the probability of observing the field in a given configuration. If the fluctuations in the field are very small, they will appear to observers as individual particles. If the fluctuations are large (which can happen only for bosonic fields as a result of the Pauli exclusion principle), they can be observed as macroscopic classical fields, such as electromagnetism and gravity.

quantum gravity: The hoped-for reconciliation of quantum mechanics and general relativity. Although the particles and forces described by the Standard Model of Particle Physics fit comfortably into the framework of quantum mechanics, it is much more difficult to describe gravity (a manifestation of the curvature of spacetime itself, according to general relativity) in quantum terms. The leading candidate for a consistent theory of quantum gravity is string theory.

quantum mechanics: The framework for physics, developed in the first half of the 20th century, in which physical systems are described by *wavefunctions* that determine the likelihood of obtaining various observational results. In quantum mechanics, as opposed to classical mechanics, the outcomes of experiments are necessarily probabilistic, rather than deterministic. A consequence of quantum mechanics is the Heisenberg uncertainty principle, which states that we cannot simultaneously fix the position and velocity of an object to arbitrary precision.

quantum number: Another name for a conserved quantity.

quark: Standard Model fermion that does feel the strong nuclear force. The six types include the up, down, charm, strange, top, and bottom quarks.

quark number: A conserved quantity given by the total number of quarks minus the total number of antiquarks. The quark number is 1/3 of the baryon number. Protons and neutrons have quark number 3; mesons have quark number 0, because they consist of one quark and one antiquark.

quintessence: A hypothetical dynamical field constituting the dark energy. Quintessence, which would feature a slowly changing energy density as the field evolves, is an alternative to vacuum energy, which would be strictly constant.

radiation: To a cosmologist, *radiation* is any collection of particles that move at or close to the speed of light. Examples would include photons (electromagnetic radiation) and gravitons (gravitational radiation). Neutrinos, which are very low mass, act as radiation in the early universe when the temperature is high but eventually cool down and behave like matter (slowly moving particles). The energy density in radiation diminishes more quickly than that in matter as the universe expands, because every individual particle loses energy due to the redshift.

radio waves: Electromagnetic waves with long wavelengths (anywhere from 1 millimeter up to 100,000 kilometers). An important example in cosmology is microwaves, with wavelengths of a few millimeters.

recombination: The moment in cosmological history when electrons combined with protons (and other nuclei) once and for all, about 380,000 years after the Big Bang. Before recombination, the universe was opaque, and afterward, it was transparent. The cosmic microwave background provides a snapshot of what conditions were like at the moment of recombination.

redshift: The stretching of the radiation from shorter to longer wavelengths as the universe expands. The cosmological redshift, caused by the expansion of space, is similar to the Doppler effect, but there are important conceptual differences.

rest energy: The energy a particle has when it is not moving. The rest energy is simply the mass of the particle times the speed of light squared ($E = mc^2$).

rotation curve: The velocity of gas and stars orbiting a galaxy, considered as a function of their distance from the galactic center. Ordinarily, the velocity of orbiting particles decreases at large distances from a gravitating object, because the force due to gravity becomes weaker. In spiral galaxies, however, the rotation speeds tend to remain constant, rather than decreasing—a phenomenon known as *flat rotation curves*. This is taken as evidence that there is additional unseen mass in the galaxies, interpreted as dark matter.

scale factor: The quantity that characterizes the relative size of the universe as it expands, often denoted $a(t)$. If the scale factor doubles in size, it means that the distance between galaxies has grown by a factor of 2.

selectron: Hypothetical supersymmetric partner of the electron. Given that the electron is a fermion, the selectron is a boson.

singularity: A point of infinite spacetime curvature. In general relativity, singularities appear inside black holes and at the Big Bang or Big Crunch. Many physicists believe that a true theory of quantum gravity will resolve such singularities and that the real world is completely nonsingular.

slepton: Hypothetical supersymmetric partner of the lepton. Given that leptons are fermions, sleptons are bosons. Each type of lepton has its corresponding slepton; for example, the superpartner of the tau is the stau.

spacetime: The four-dimensional manifold describing the universe. Relativity took the separate Newtonian concepts of three-dimensional absolute space and one-dimensional absolute time and replaced them with four-dimensional spacetime. The division of spacetime into "space" and "time" will not look the same to all observers. In general relativity, spacetime can be curved, and that curvature manifests itself as gravity.

spatial curvature: See **curvature of space**.

special relativity (SR): Einstein's theory of the unification of space and time into a single concept of spacetime. It is based on the idea that the speed of light is an absolute maximum velocity that looks the same to all observers. According to SR, the elapsed time along a trajectory is not the same for all paths connecting two events but depends on the route taken through spacetime.

speed of light: Usually denoted c, the speed of light in a vacuum is the maximum velocity at which two objects can pass by each other. (In general relativity, the relative velocity of two distant objects is not well defined.) It is approximately equal to 300,000 kilometers per second. Light travels at slower speeds when it moves through a medium such as air or glass; however, the phrase *speed of light* generally refers to the speed in a vacuum.

spin: An intrinsic amount of angular momentum generally characteristic of elementary particles.

squark: Hypothetical supersymmetric partner of the quark. Given that quarks are fermions, squarks are bosons. Each type of quark has its corresponding squark; for example, the superpartner of the top quark is the stop squark.

stable: Particles are stable if they would last forever if left by themselves.

standard candles: Objects of known brightness at astronomical distances; useful as distance indicators. If we know how bright an object really is and how bright it appears we can figure out how far away it must be.

Standard Model of Particle Physics: The modern theory of elementary particles and their interactions. The Standard Model includes 12 kinds of fermions—6 kinds of leptons and 6 kinds of quarks—and 3 types of gauge bosons that mediate forces between them. It also includes the Higgs boson, which has not yet been discovered. It does not, strictly speaking, include gravity, although it's often convenient to include gravitons among the known particles of nature.

strange quark: An elementary particle; one of the six types of quarks. It is somewhat heavier than the up and down quarks and, therefore, plays only a minor role in ordinary matter. To be honest, not that strange, as quarks go.

string theory: The idea that the elementary constituents of matter are small loops of string rather than pointlike particles. String theory was originally suggested as a theory of the strong interactions (before QCD was invented) but inevitably predicted the existence of a particle with all the characteristics of the graviton. These days, we think of string theory as a candidate theory of quantum gravity. Unfortunately, it has proven difficult to derive explicit testable predictions from the theory that can be compared with experiment.

strong lensing: Gravitational lensing that is strong enough to create multiple images of lensed objects.

strong nuclear force: The force that binds quarks together in hadrons. It is mediated by gluons, as described by quantum chromodynamics. The strong force between two quarks actually grows stronger as the quarks are pulled apart, which leads to the effect of confinement.

Sunyaev-Zeldovich (SZ) effect: When low-energy photons from the cosmic microwave background pass through the hot gas in a cluster of galaxies, they can be scattered to higher energies. This is the *Sunyaev-Zeldovich effect*, which can be observed by mapping the energies of CMB photons in the direction of clusters on the sky. The SZ effect is similar to a shadow being cast on the CMB.

supernova (pl.: supernovae): An exploding star. A particular type of supernova, known as *Type Ia*, plays an important role in cosmology because it is a standardizable candle. Type Ia supernovae arise when enough mass accretes onto a white dwarf star to pass the Chandrasekhar limit. The white dwarf then collapses, and the outer layers are expelled in a violent explosion, almost as bright as an entire galaxy.

superpartner: In supersymmetry, the hypothetical particles associated with the particles that are already known in the Standard Model; superpartners of fermions are bosons, and superpartners of bosons are fermions. For example, the superpartner of the electron is the selectron, and the superpartner of the photon is the photino.

superstring: Strings (as in string theory) that are also supersymmetric. It is widely believed that string theory must be supersymmetric to describe the real world.

supersymmetry: A hypothetical symmetry relating bosons to fermions. If supersymmetry exists in the real world, it must be broken, so that the partners of the known Standard Model particles are too heavy to have yet been discovered. The lightest such particle, if it's neutral and stable, makes an excellent dark matter candidate.

symmetry breaking: The situation that is said to exist when the symmetry of a theory is hidden from view by the dynamics of a system. In the electroweak sector of the Standard Model of Particle Physics, the Higgs field acts to break the electroweak symmetry. If supersymmetry is part of the real world, some unknown mechanism must act to break it as well, because superpartners with the same mass as ordinary particles are not observed in nature.

tau: An elementary particle; the heaviest charged lepton. The tau has a very short lifetime, as it rapidly decays into lighter leptons.

tau neutrino: An elementary particle; the neutrino that interacts directly with the tau.

temperature: A measure of the average kinetic energy of particles in a system. As the universe expands, the temperature tends to decrease as the reciprocal of the scale factor.

tension: Negative pressure. If a substance pulls things together, like a stretched rubber band, it has a tension. Dark energy has tension but (apparently paradoxically) makes galaxies accelerate away from each other. That's because the important effect is not directly from the tension but through its gravitational influence. The direct pull of the tension is exactly the same in every direction and, therefore, cancels out.

top quark: An elementary particle; one of the six types of quarks. The top is the heaviest quark.

unstable: Particles that decay when left by themselves are unstable.

up quark: An elementary particle; one of the six types of quarks. It is the lightest quark and, as a constituent of both protons and neutrons, plays an important role in ordinary matter.

vacuum: The most common meaning of *vacuum* is simply "empty space." It can also refer to the lowest energy state of a physical system.

vacuum energy: The energy density of empty space, also known as the cosmological constant. The vacuum energy is not caused by a dynamical field but is a property of spacetime itself—the minimum amount of energy that can possibly exist in some region of space. It is, therefore, constant throughout space and time and, in particular, does not diminish as the universe expands. Vacuum energy is a leading candidate for the dark energy.

vacuum fluctuations: According to the Heisenberg uncertainty principle, quantum fields will inevitably fluctuate, even in empty space. These vacuum fluctuations take the form of virtual particles rapidly popping in and out of existence. The energy of the vacuum fluctuations is an important contributor to the vacuum energy.

virtual particle: A particle that briefly fluctuates into existence, then disappears as a result of the inherent uncertainty in quantum mechanics.

W boson: Charged force-carrying particles of the weak interactions. Unlike the photon (which carries the electromagnetic force), W bosons are very massive, explaining why the weak force operates only over short ranges.

wavefunction: The function in quantum mechanics that encodes the state of a physical system. The wavefunction provides the information necessary to calculate the probability of obtaining any specific answer when the system is observed—for example, the position or velocity of a particle. It is impossible to observe a generic wavefunction without changing the quantum state of the system; this is the origin of the Heisenberg uncertainty principle.

weak lensing: Gravitational lensing that is strong enough to distort images of objects but not strong enough to break them into multiple images. Given that we usually don't know the original shape of the images, the effects of weak lensing can only measured statistically, by observing many objects in one region.

weak nuclear force: The short-range, relatively feeble force carried by the W and Z bosons. The weak force is responsible for certain important radioactive decays, such as the decay of the neutron. At high energies, it combines with electromagnetism to form the electroweak interaction.

weakly interacting massive particle (WIMP): A candidate particle for dark matter. The phrase *weakly interacting* in this context doesn't simply mean that the particle doesn't interact very much; it refers specifically to the weak nuclear force of the Standard Model. Massive, stable particles that interact through the weak force and not through the strong force or electromagnetism turn out to naturally have an appropriate relic density to make up the dark matter. A good example of a WIMP candidate is the lightest supersymmetric particle (LSP).

white dwarf: A star that has used up all its nuclear fuel. White dwarfs are supported by the pressure from their electrons, which cannot be squeezed too tightly because of the Pauli exclusion principle.

X-rays: High-energy electromagnetic radiation, with wavelengths between 10^{-11} and 10^{-8} meters.

Z boson: Neutral force-carrying particle of the weak interactions. Unlike the photon (which carries the electromagnetic force), Z bosons are very massive, explaining why the weak force operates only over short ranges.

Biographical Notes

Einstein, Albert (1879–1955). German-American physicist. Einstein is, of course, the one physicist everyone can name. His fame is well deserved; when he was chosen as *Time* magazine's "Person of the Century," it was hard to argue with the choice.

In 1905 Einstein was two years out of school and working at the patent office in Zurich, unable to find an academic job. That year, he published a series of papers that revolutionized modern physics, among which was the first comprehensive statement of the principles of special relativity. This theory replaced Newtonian mechanics and fit elegantly with Maxwell's laws of electricity and magnetism. It was not, however, compatible with Newton's theory of gravity. For the next 10 years, Einstein worked to reconcile gravity with relativity, culminating in the formulation of general relativity in 1915. In 1919, a British team led by Arthur Eddington took observations during a total eclipse that confirmed general relativity's prediction of light-bending due to the gravitational field of the Sun, making Einstein an international celebrity.

Einstein was never comfortable with the apparent loss of determinism implied by quantum mechanics (even though some of his work helped found the field). In his later years, he worked on an ultimately unsuccessful program to unify gravitation and electromagnetism; this attempt caused him to drift away from the mainstream of physics at the time, but it foreshadowed later successes at unifying the forces. In 1932, just before Hitler was elected chancellor of Germany, Einstein moved permanently to the United States. He became active in politics, helping to urge President Franklin Roosevelt to develop the atomic bomb and, in 1952, declining an offer to become Israel's second president.

Freedman, Wendy (b. 1957). American astronomer. First to accurately measure the Hubble constant. Freedman received her Ph.D. in 1984 from the University of Toronto. Since then, she has worked at the Carnegie Observatories in Pasadena, California, where she now serves as director. She was the principal investigator for one of the primary missions of the Hubble Space Telescope Key Project to determine the Hubble constant. For decades after Hubble's original discovery, astronomers had struggled to determine the numerical value of the constant relating the distance of a galaxy to its apparent recession velocity; it's this number that sets the scale for the size and age of the observable universe. Freedman and her collaborators were able to establish that the Hubble constant is approximately 72 kilometers per second per megaparsec: For each megaparsec of distance, a galaxy will be receding at 72 km/sec.

Friedmann, Alexander (1888–1925). Russian physicist. Friedmann was the first person to systematically study cosmology in the context of general relativity and, in 1922, derived the Friedmann equation for the evolution of the universe. Like many famous figures in the history of general relativity, he fought in World War I. Friedmann died young, of typhoid fever. One of his students, George Gamow, became an influential theoretical physicist and cosmologist.

Galilei, Galileo (1564–1642). Italian physicist. The first person to understand conservation of momentum and universality of free fall. Also the first to use a telescope for astronomical observations and provide evidence that the Earth moves around the Sun. Galileo, it is safe to say, was at least the grandfather of modern physics (if Newton was the father).

Born in Pisa, Galileo briefly studied medicine before switching to mathematics, in which he became a lecturer at the University of Pisa in 1589; in 1592, he moved to the University of Padua. He was an extraordinary source of new ideas and experiments. A short list of his accomplishments in physics would include the discovery that the period of a pendulum is independent of its amplitude, a demonstration that the distance an object falls is proportional to the square of the time of its fall, an attempted measurement of the speed of light, a demonstration that objects of different compositions fall at the same rate, and (most importantly) the realization that moving objects which are not acted on by any force will retain their velocity. In 1609, Galileo built his first telescope and immediately turned it on the night sky. He found the first evidence for the rings of Saturn; showed that Venus had phases, just as the Moon does; and discovered the satellites of Jupiter. This last finding was the first example of celestial bodies that clearly orbited an object other than the Earth, calling into question the geocentric Ptolemaic system.

In 1611, Galileo visited Rome to demonstrate some of his discoveries to Church authorities. Although initially supportive, the

Church ultimately asked him to treat the heliocentric model of Copernicus as merely a useful hypothesis rather than as a true representation of reality. But in 1623, Galileo's friend Cardinal Barberini was elected pope, which encouraged Galileo to publish a book on the controversy. In these dialogues, he gave the job of defending geocentrism to the character Simplicio, which the pope took as public ridicule. In 1633, Galileo was convicted of heresy and put under house arrest, where he stayed until his death. He had two daughters and one son out of wedlock; the daughters were sent to a convent in Arcetri, where they remained for their entire lives.

Gamow, George (1904–1968). Ukranian-American physicist. Gamow is responsible for much of our modern understanding of the early universe in the context of the Big Bang model. His early career took him from Odessa (Ukraine) to Leningrad (Russia) to Göttingen (Germany) to Copenhagen (Denmark) to Cambridge (England), before returning to the Soviet Union. His student years coincided with the birth of quantum mechanics, and Gamow made a significant contribution to the field, being the first to understand the process of radioactive decay as the result of quantum tunneling.

In 1933, Gamow and his wife defected from the USSR, and in 1934, they moved to the United States for good. With his students Ralph Alpher and Robert Herman, Gamow wrote a series of papers in the 1940s and 1950s that laid out the early history of the universe in the Big Bang theory. These included the process of Big Bang nucleosynthesis and the formation of the cosmic microwave background, including a rough prediction of its temperature. Later in life, Gamow made important contributions to the theory of DNA and wrote a series of extremely successful popular books on science, including *One, Two, Three...Infinity* and the *Mr. Tompkins* series.

Guth, Alan (b. 1947). American physicist. Inventor of the theory of cosmic inflation. Born in New Jersey, Guth skipped his senior year of high school to attend the Massachusetts Institute of Technology. He stayed on at MIT for his doctorate, working on the theory of quarks. After getting his Ph.D., Guth took on a series of postdocs; he was unable to land a faculty position because he hadn't published many papers. But he gradually became interested in questions of cosmology and, in 1980, had a spectacular realization: Many problems of the conventional cosmological model could be solved in a single swoop by postulating a period of accelerated expansion in the first fraction of a second in the history of the universe. His theory of cosmic inflation caused a sensation, and he was swiftly offered a professorship back at MIT, where he is still teaching today. The research notebook recording the flash of insight in which Guth first understood the implications of inflation is currently on display at the Adler Planetarium in Chicago.

Hubble, Edwin (1889–1953). American astronomer. Showed that many nebulae were galaxies in their own right ("the universe is big") and established that the universe is expanding ("and it's getting bigger"). Hubble graduated with a bachelor's degree in 1910 from the University of Chicago, where he stood out more as an athlete than as a student. He then went to Oxford as a Rhodes Scholar, where he studied jurisprudence and Spanish. Returning to the United States, he worked as a lawyer and as a high school teacher and basketball coach before serving in World War I. After the war, he returned to the University of Chicago, earning a Ph.D. in astronomy in 1917.

Hubble was offered a position at the newly created Mount Wilson Observatory in California, where he stayed for the rest of his life. Using the 100-inch Hooker telescope, by 1925, he was able to identify individual pulsating Cepheid stars in Andromeda and other "nebulae," proving that they were actually galaxies like the Milky Way. Hubble and Milton Humason then combined their measurements of distances with redshift data obtained by Vesto Slipher to establish that more distant galaxies appeared to be receding more rapidly. This result, announced in 1929, can be interpreted in the context of general relativity as arising from the expansion of the universe. Hubble never won the Nobel Prize; during his lifetime, astronomy was not considered to fall under the category of physics for purposes of the prize.

Lemaître, Georges-Henri (1894–1966). Belgian priest and physicist. Lemaître was the first person to take seriously the initial cosmological singularity—what we now call the Big Bang, which he referred to as the "Cosmic Egg" or the "Primeval Atom." Friedmann was the first to derive the equations governing the expansion of the universe in general relativity, but Lemaître's 1927 paper derived Hubble's law on theoretical grounds and emphasized the singular beginning of the universe. It wasn't until 1929 that Hubble and Humason would demonstrate that the universe really is expanding.

Lemaître led a colorful life. His studies were interrupted so that he could serve in the Belgian army in World War I. Afterward, he simultaneously studied physics and prepared for the priesthood. His studies took him to the University of Cambridge (England), Harvard, and the Massachusetts Institute of Technology. After Hubble and Humason's discovery, Lemaître popularized his Primeval Atom theory in lectures in Britain and the United States and became a well-known figure. Later in life, he became president of the Pontifical Academy of Sciences and served on the Second Vatican Council.

Mather, John (b. 1946). American astrophysicist. Performed the first precision measurement of the blackbody nature of the cosmic microwave background. Mather was an undergraduate at Swarthmore College and received his Ph.D. from Berkeley. He worked for a while at Columbia University before moving to NASA's Goddard Space Flight Center. It was at Columbia, in 1974, that he first began working on the plans for what would eventually become the Cosmic Background Explorer (COBE) satellite, which was ultimately launched in 1989. Mather was the lead investigator for the mission as a whole and for the Far Infrared Spectrophotometer (FIRAS) instrument, which measured the CMB spectrum. He shared the Nobel Prize in Physics in 2006 with George Smoot.

Newton, Isaac (1643–1727). British physicist. Inventor of classical mechanics, calculus, and the inverse-square law of gravity and, perhaps, the greatest scientist of all time. Newton attended Cambridge University, where he became fascinated with the modern ideas of such thinkers as Descartes and Galileo. Several of Newton's discoveries—such as the inverse-square law of gravity, a design for a new kind of telescope, and the separation of white light into its component colors using a prism—would have individually been enough to grant him scientific immortality. But they pale in comparison to his invention of calculus—arguably the most important advance in the history of mathematics—and to his formulation of classical mechanics and the laws of motion—arguably the most important advance in the history of physics. These accomplishments were not without controversy; calculus, in particular, was the subject of a decades-long priority dispute with German mathematician Gottfried Leibniz. Newton probably did his work first but didn't publish until after Leibniz.

Classical mechanics, first published as part of Newton's *Principia Mathematica*, formed a basis for physics that is still taught to students today. It consists of Newton's three laws of motion: Objects in uniform motion stay in uniform motion unless acted on by an outside force; force equals mass times acceleration; and for every action, there is an equal and opposite reaction. These rules were ultimately superseded by the laws of relativity and quantum mechanics but were precise enough to land astronauts on the Moon. To gauge the scale of Newton's accomplishment, in *Blind Watchers of the Sky*, Rocky Kolb compares it to the Wright brothers putting together a modern jumbo jet and flying it from Kitty Hawk to New York. Together with his law of gravity, Newton was able to use the laws of classical mechanics to derive Kepler's laws of planetary motion.

Newton never married and is thought to have died a virgin. Along with his scientific studies, he managed to have a career in public life, serving briefly as a member of Parliament and for an extended time as master of the Royal Mint. He used this latter position to persecute his enemies, of which he had many. He also moved England to the gold standard in 1717. He performed considerable experiments in alchemy and spent much of his later years publishing extensively on religion.

Penzias, Arno (b. 1933). American astrophysicist. Co-discoverer (with Robert Wilson) of the cosmic microwave background. Penzias was born in Germany; like many other Jewish families, he and his parents fled when Hitler came to power, and he became an American citizen in 1946. After getting his Ph.D. from Columbia, he moved to Bell Labs in New Jersey, where he worked on the emerging field of radio astronomy. Penzias and Wilson constructed an ultra-sensitive radiometer but encountered a mysterious source of background noise that they could not account for. They ultimately realized that it was cosmic in origin and learned that a group in nearby Princeton was searching for exactly this radiation. Penzias and Wilson shared the Nobel Prize in 1978 with Pyotr Kapitsa, a Russian physicist who discovered the phenomenon of superfluidity.

Perlmutter, Saul (b. 1959). American astronomer. Leader of the Supernova Cosmology Project, one of the two teams to discover the acceleration of the universe. Perlmutter was an undergraduate at Harvard and earned his Ph.D. at Berkeley under the supervision of

Richard Muller. His thesis project centered on the development of a robotic telescope that searched for supernovae. After it became clear that Type Ia supernovae could be used as distance indicators, Perlmutter and Carl Pennypacker launched a project to search for supernovae in wide-field images of thousands of galaxies at a time. This effort grew into the Supernova Cosmology Project, which by 1994 was gathering supernovae by the bunches. Perlmutter was awarded the Shaw Prize along with Adam Riess and Brian Schmidt in 2006 for their discovery of the acceleration of the universe. Perlmutter is currently a scientist at Lawrence Berkeley Laboratory and a professor at the University of California, Berkley.

Riess, Adam (b. 1969). American astronomer. Member of the High-Redshift Supernova Team, one of the two teams to discover the acceleration of the universe. Born in Washington, DC, Riess was an undergraduate at MIT before moving to Harvard for graduate school. Like Brian Schmidt, he worked with Robert Kirshner on using supernovae to measure the Hubble constant; his Ph.D. thesis received the Trumpler Award for the most influential dissertation in astrophysics. While working as a postdoc at Berkeley, Riess was the lead author on the first paper to announce that the universe is accelerating. He was awarded the Shaw Prize along with Brian Schmidt and Saul Perlmutter in 2006 for their discovery of the acceleration of the universe. He is currently a scientist at the Space Telescope Science Institute in Baltimore.

Rubin, Vera (b. 1928). American astronomer. Discovered dark matter in spiral galaxies. When she applied to Swarthmore as a high school student intending to study science, she was told by an admissions officer that she might be better off pursuing her interests in art, perhaps by painting pictures of astronomical objects; she chose to attend Vassar instead. Rubin tried to attend Princeton for her master's degree, but Princeton didn't admit women into the graduate school at the time, so she went to Cornell instead. In 1954, she obtained a Ph.D. from Georgetown, where George Gamow was her supervisor. In 2005, Princeton eventually made up for its mistake, awarding her an honorary doctorate, which she has also received from Harvard and Yale.

After graduation, Rubin continued on at Georgetown as a researcher and faculty member, before moving to the Department of Terrestrial Magnetism at the Carnegie Institution in 1965. In the 1970s, she undertook an ambitious program, frequently in collaboration with Kent Ford, to study the rotation of nearby spiral galaxies. Rubin demonstrated conclusively that gas in the outer regions of galaxies was moving much faster than could be accounted for by the gravitational field of the visible matter alone, providing strong evidence for the existence of dark matter. She was awarded the National Medal of Science in 1993.

Schmidt, Brian (b. 1967). American-Australian astronomer. Leader of the High-Redshift Supernova Team, one of the two teams to discover the acceleration of the universe. After growing up in Montana and Alaska, Schmidt attended Harvard for his graduate work, where he wrote a thesis on using supernovae to measure the Hubble constant under the supervision of Robert Kirshner. After moving to Mount Stromlo and Siding Springs Observatories in Australia, Schmidt put together a team to discover large numbers of supernovae and use them to learn about cosmology. He was awarded the Shaw Prize along with Adam Riess and Saul Perlmutter in 2006 for their discovery of the acceleration of the universe.

(Full disclosure: Your humble lecturer was officemates with Schmidt in graduate school and can vouch that Brian is a gourmet cook as well as a world-class astronomer.)

Smoot, George (b. 1945). American astrophysicist. Discoverer of temperature anisotropies in the cosmic microwave background. Smoot obtained both his undergraduate and graduate degrees at MIT. Originally interested in particle physics, he switched to cosmology, perceiving it to be a less crowded field. After moving to Berkeley, he began to study the microwave background. In 1977, he discovered a slight anisotropy in the temperature of the CMB, but this was due to the motion of the Milky Way galaxy through space, not to primordial fluctuations. Smoot was the lead investigator on the Differential Microwave Radiometer (DMR) experiment on the Cosmic Background Explorer (COBE) satellite, launched in 1989. It was this project that, in 1992, announced the discovery of primordial temperature fluctuations from which large-scale structure in the universe ultimately evolved. In 2006, Smoot shared the Nobel Prize in Physics with John Mather, leader of the COBE team.

Wilson, Robert (b. 1936). American astrophysicist. Co-discoverer (with Arno Penzias) of the cosmic microwave background. Wilson was an undergraduate at Rice University and received his first patent the summer after he graduated, before moving to Caltech for his Ph.D. work. In 1963, he joined Bell Labs in New Jersey, where he began collaborating with Penzias on radio instrumentation. After the discovery of the CMB in 1965, Wilson continued to work on radio astronomy, mapping out organic molecules in the Milky Way. Penzias and Wilson shared the Nobel Prize in 1978 with Pyotr Kapitsa, a Russian physicist who discovered the phenomenon of superfluidity.

Zwicky, Fritz (1898–1974). Swiss-American astrophysicist. First to propose the concept of dark matter. Born in Bulgaria to Swiss parents and educated in Switzerland, Zwicky moved to the United States in 1925 to take a position at Caltech, where he remained for the rest of his life. In 1933, his analysis of the dynamics of the Coma cluster of galaxies led to the surprising discovery that there was much more matter in the cluster than implied by the visible gas and stars. Zwicky and Walter Baade coined the term *supernova*, and Zwicky was personally responsible for the discovery of more than 100 supernovae. He also did pioneering early work on the theory of neutron stars, cosmic rays, and gravitational lenses.

Bibliography

A large number of great books have been written about cosmology, particle physics, and gravity. None of them follows this course precisely, especially because these are fast-moving fields in which discoveries keep coming at us unexpectedly. But all the books listed here are good reads at an accessible level and will help flesh out the ideas discussed in the lectures.

Recommended Readings:

Cole, K. C. *The Hole in the Universe: How Scientists Peered over the Edge of Emptiness and Found Everything.* San Diego, CA: Harcourt, 2001. A far-ranging look at the concept of "nothing" in physics, including the energy of empty space.

Ferris, Timothy. *The Whole Shebang.* New York: Simon and Schuster, 1997. An in-depth account of modern cosmology, its practitioners, and its implications.

Feynman, Richard. *The Character of Physical Law.* Cambridge, MA: MIT Press, 1965. A brief introduction to the way the laws of physics work, by someone who understood them as well as anyone.

Freeman, Ken, and Geoff McNamara. *In Search of Dark Matter.* New York: Springer Praxis Books, 2006. A nice, short overview of the dark matter problem. Explains in detail how we weigh galaxies and how we distinguish ordinary baryonic matter from exotic dark matter.

Goldsmith, Donald. *The Runaway Universe: The Race to Find the Future of the Cosmos.* Cambridge, MA: Perseus, 2000. A brisk report of the 1998 discovery of the accelerating universe by two supernova cosmology teams and its aftermath.

Greene, Brian. *The Fabric of the Cosmos.* New York: Knopf, 2004. A modern survey of space, time, and gravity, from Einstein's relativity to modern cosmology and superstring theory.

Guth, Alan. *The Inflationary Universe.* Cambridge, MA: Perseus, 1998. A masterful account of modern cosmology, especially the early universe, from the inventor of the inflationary universe scenario.

Hogan, Craig. *The Little Book of the Big Bang.* New York: Copernicus, 1998. Just what the title promises—a short and engaging survey of the basic picture of Big Bang cosmology.

Kirshner, Robert. *The Extravagant Universe: Exploding Stars, Dark Energy, and the Accelerating Cosmos.* Princeton, NJ: Princeton University Press, 2002. A first-person account of one astronomer's research on supernovae, culminating in the discovery that the universe is accelerating.

Kolb, Rocky. *Blind Watchers of the Sky.* Reading, MA: Addison-Wesley, 1996. A gripping history of cosmology, told through biographical sketches of its leading lights.

Lemonick, Michael. *Echo of the Big Bang.* Princeton, NJ: Princeton University Press, 2003. A behind-the-scenes look at the development and launch of the WMAP satellite and what we learned from its observations of the cosmic microwave background.

Levin, Janna. *How the Universe Got Its Spots.* Princeton, NJ: Princeton University Press, 2002. A highly personal and fascinating story of one young researcher's investigations into black holes and the topology of the universe.

Livio, Mario. *The Accelerating Universe: Infinite Expansion, the Cosmological Constant, and the Beauty of the Cosmos.* New York: Wiley, 2000. A short book on the beauty of modern cosmology and its connections to the arts.

Nicolson, Ian. *The Dark Side of the Universe: Dark Matter, Dark Energy, and the Fate of the Cosmos.* Baltimore, MD: Johns Hopkins University Press, 2007. A recent book that focuses on dark energy; includes a large number of color photos.

Osserman, Robert. *Poetry of the Universe.* Garden City, NJ: Anchor Books, 1995. An accessible introduction to cosmology from the perspective of a mathematician.

Overbye, Dennis. *Lonely Hearts of the Cosmos.* New York: Harper Collins, 1991. The story of modern cosmology, told in an engaging and personal style.

Randall, Lisa. *Warped Passages: Unraveling the Mysteries of the Universe's Hidden Dimensions.* New York: Ecco, 2005. A recent, in-depth examination of the idea of extra dimensions by one of the field's leading innovators.

Vilenkin, Alex. *Many Worlds in One: The Search for Other Universes*. New York: Hill and Wang, 2006. An investigation of the consequences of the possibility that our universe is just one of many.

Wald, Robert. *Space, Time, and Gravity*. Chicago: University of Chicago Press, 1992. A short, slightly technical introduction to Einstein's theories of special and general relativity originally published in 1977, with a revised and updated edition that appeared in 1992.

Weinberg, Steven. *Dreams of a Final Theory*. New York: Pantheon, 1992. An examination of modern particle physics, with an emphasis on the role of fundamental theory and the virtues of reductionism.

―――. *The First Three Minutes*. New York: Basic Books, 1977. Slightly out of date but still a classic exposition of early-universe cosmology by a Nobel laureate.

Will, Clifford. *Was Einstein Right?* New York: Basic Books, 1993. A valuable look at alternatives to Einstein's theory of general relativity and the experimental tests to which modern physics subjects them.

Internet Resources (Web sites for general tutorials, as well as updated news items):

American Institute of Physics. *Cosmic Journey*. A well-constructed site that provides an introduction to modern cosmology, as well as a historical survey. There are also discussions of the instruments used by astronomers from early times to the present. www.aip.org/history/cosmology/.

Carroll, Sean. *Cosmology Primer*. A brief overview of the basics of modern cosmology, from observations of large-scale structure and the microwave background to the very early universe. preposterousuniverse.com/writings/cosmologyprimer/.

Contemporary Physics Education Project. *The Universe Adventure*. A companion to *The Particle Adventure* (below), this interactive site introduces the basic ideas of Big Bang cosmology at an accessible level. www.cpepweb.org/main_universe/universe.html.

Cosmic Variance Blog. An informal Web log with Sean Carroll and several other contributors. All the authors are physicists and astronomers, and conversations range from science news to whatever else crosses their minds. cosmicvariance.com/.

Feuerbacher, Björn, and Ryan Scranton. *Evidence for the Big Bang.* A highly detailed but accessible overview of the many pieces of evidence we have in favor of the basic Big Bang picture. Includes detailed refutations of many common skeptical arguments. www.talkorigins.org/faqs/astronomy/bigbang.html.

Lawrence Berkeley National Laboratory. *The Particle Adventure.* An interactive site that provides a basic introduction to particle physics. A detailed tutorial is accompanied by numerous links to associated resources. particleadventure.org/.

NASA. *Cosmology 101.* A site associated with the WMAP cosmic microwave background satellite but also including an introduction to basic cosmology. map.gsfc.nasa.gov/m_uni.html.

Space Telescope Science Institute. *The Hubble Site.* A public site associated with the Hubble Space Telescope (HST). Includes news about astronomy in general and HST in particular, as well as a breathtaking image gallery and explanations of Hubble's many breakthroughs. hubblesite.org/.

Wright, Edward L. *Ned Wright's Cosmology Tutorial.* A tutorial at a slightly higher level than many of the previous sites. Uses basic algebra, as well as detailed presentations of modern astronomical data. www.astro.ucla.edu/ ~wright/cosmolog.htm.

Notes

Notes